South Dakota's Fairburn Agate

by Roger Clark

photography by Mary Jane Clark

First Edition

ISBN Number: 0-9664640-0-1

Library of Congress Number: 98-090403

Published by Silverwind Agates
800 North Lynndale Drive
Appleton, WI 54914

Printed in the United States of America
by Palmer Publications, Inc.
318 North Main Street
Amherst, WI 54406

Front Cover:
Fairburn agates from the author's collection
which have not been cut or polished.

Dedication

To my wife, Mary Jane,
who is my photographer,
traveling companion, and best friend.

Contents

Acknowledgements

The author gratefully acknowledges Art and Ann Bruce, George Burg and Jack Seger for sharing their collections to be photographed and their experiences to contribute to this writing.

John Paul Gries was invaluable for his knowledge of the Geology of the Black Hills and surrounding areas.

My staff people, Melissa Meirhofer and Dixie Thoyre, were always patient and supportive as the work progressed, and for tending to business while I explored South Dakota.

Project Coordinator: Marcia Lorenzen
Project Design: Heidi Bittner-Zastrow
Text Editor: Terrell W. Bonnell

Introduction

The Fairburn agate is named for the small community of Fairburn, South Dakota. The agate is known for its spectacular variety of color and the fortifications which in many instances occur in a "holly leaf" type of pattern. According to the late Jack Zasadil in his article in *The Agates of North America* published by The Lapidary Journal (1966) they are found...

> *...from Scenic, to the northeast of Hermosa, to Orella, Nebraska, just across the South Dakota line. I have heard of them being found from as far south as Gordon, Nebraska, and from my place at Hermosa eastward to Cooney (sp. Cuny) Table, thirty miles away.*

This book will explain why the author believes Fairburns are found in an even broader range than described by Jack Zasadil.

Fairburn hunters are blessed with huge areas of National Grasslands and Black Hills National Forest lands that are open to all who want to explore. Grassland areas where significant erosion has occurred are often referred to as badlands. These references should not be confused with The Badlands National Park. No rock hunting is allowed in the national park.

When westward travelers stop at Wall Drug, they are stopping at the gateway to the entire corner of the state, extending south to Nebraska and west to Wyoming. Fairburns may be found anywhere in that area.

The book is not designed to be a trail or guide book. Still the best analysis and hunting tool available is June Zeitner's *Midwest Gem Trails*. If you are interested in actual hunting, it is an excellent place to start. You can also get a firsthand look at her Fairburn collection, which is on display at the Pioneer Auto Show in Murdo, South Dakota. The Auto Show buildings also house the National Rock Hound Hall of Fame.

Contacts with local rock clubs will usually provide the names of people willing to share their rock hunting experiences.

Most of the photographs in this book do not have a reference for perspective (size of specimen). Color and pattern were the most important to this work. When there is reference to size, you will see the terms small (up to two inches), medium (two to four inches), and large (over four inches). As a general rule, the best color and patterns are found in the small and medium agates.

Chapter 1

South Dakota's Fairburn; beginning the exploration.

A Personal Journey

This book is a personal story of exploration emerging out of an acquired passion to learn about South Dakota's state gemstone, the Fairburn agate. The search for the origin of the agate in South Dakota also led me on a search to understand the origin or genesis of agates. There will be readers of this book who have substantial knowledge about the geology of South Dakota's Black Hills and Badlands. Those readers will have to appreciate that I have taken some liberty in generalizations of geologic history. This has been done to make the book more friendly as opposed to a technical book. Hopefully, the majority of readers will welcome some new knowledge and information about South Dakota's state gemstone.

A comprehensive general study of agate formation was reviewed from true professionals in the field such as Landmesser, Pabian and Shaub. These are men who have spent significant parts of their lives working with geology, mineralogy, and chemistry. The actual debate about the genesis of agates, in particular the varying types of agates, including banded agates, still continues. The reader must also understand that the theories of the chemistry and mineralogy are complex. The summaries set forth in this text are the interpretation of the author. My attempt at brevity may not do justice to the works of Landmesser and Pabian.

Had I been born in South Dakota, I probably would have been agate hunting for 40 years instead of 22. Since I am not a native of the state, there will be questions about the writings by an "out-of-state" Fairburn hunter. The questions such as: What does he know of the old-time Fairburn hunting? What does he know of the grasslands, badlands, and the Black Hills? What kind of collection could he have, especially when Fairburns are so scarce? Who does he think he is proposing that he has established the origin of the Fairburns???

This writing grew out of my love for South Dakota's Fairburn agate. It is the summary of my knowledge gained not only from hunting, but also from those people I consider the real experts; old-time hunters like Art and Ann Bruce, George Burg, Bill Zieg; new-age hunters like Bill Stein and a group of rock hunting friends from Rapid City and Hot Springs. It contains geology information and geology input from the School of Mines and the likes of J. Paul Gries, a long-time professor at the School of Mines and author of the *Roadside Geology of South Dakota*. There is also my accumulation over 22 years, of a personal collection of Fairburns both purchased and collected from the grasslands and badlands. There is an additional collection of samples of Fairburn agates from the Black Hills. (Yes, I said Black Hills.)

This writing proposes some new thoughts and ideas based upon research and observations. New ideas that will hopefully inspire agate hunters from South Dakota and from all over to bring forward additional information about anything set forth in the writing.

Where are the Fairburns?

One of my frustrations as a rock hound has been that the Fairburns seem to have fallen into collectors' basements where, in most instances, they remain unseen. I am told about large collections of beautiful agates, but they are seldom displayed in public except for the museum at the School of Mines. Gries advises me that the School of Mines collection of rocks and minerals, in part, came from the Depression when some students were allowed to use their rocks and fossils to pay their tuition to school. The Western Dakota Gem and Mineral Society at the Rapid City Show, does bring out some of the collections for a brief two-day display. The Pioneer Auto Show in Murdo, South Dakota, also houses the Zeitner collection. There are some nice Fairburns on display.

Fairburns can be seen displayed in the rock shops of southwestern South Dakota (when traveling west) beginning with the rock shop at Wall Drug. Reasonably priced small samples can be purchased, but Fairburns with good color and pattern are expensive.

To gain an appreciation of the Fairburn agate, you need to be able to see some of them. It was the "showing" of a representation of the agates that was a motivation for the photography in this book.

Other than Zeitner's *Midwest Gem Trails*, information on collecting is scarce. In the late 1970s, I heard of a Fairburn auction in Hot Springs, South Dakota. The information came to me after the fact. Because I wanted to know who would be putting Fairburns up for auction, I tracked down the auc-

The original Fairburn beds demonstrate the eroding grasslands which created badlands that uncover more rock with each passing year.

tioneer and got the seller's name. The name was given to me as Art Bruce, who lived in Hot Springs. After meeting Art and Ann Bruce, I was amused when a person I met in a local rock shop told me he knew of a local collector that not only did not show his agates...but slept with them under his bed. That anecdote was a good indication that Fairburn collectors were very cautious about publicity. Art and Ann Bruce used to keep Fairburns under their bed.

If this book will raise enough controversy to bring Fairburn collectors out in the open, it will be successful. I would be very interested in having people come forward and tell me about their collections or the collections of others. Disclosure of secret locations may help to substantiate the theories in this writing. There are a large number of collectors and hunters that I have not yet had the privilege to meet. I would be pleased to be told about old-time finds that could support or disprove, or shed new light on the origin and history of Fairburn agates.

A long shadow on an early morning visit to the "Nebraska" agate beds. Rocks like these are scattered throughout the grasslands. These beds are the remains of alluvial deposits from the erosion of South Dakota's Black Hills.

As a caution to readers who may want to hunt, you must know that there is no rock collecting permitted in the Badlands National Park. The park area is often referred to as the "Big" Badlands. Throughout this book, you will see reference to the "badlands". This term encompasses the badlands that are **not in the park** which also include prairies that are the Buffalo Gap National Grasslands, and Oglala National Grasslands (Nebraska). Because of the erosion that occurs along the streams and rivers, there are many areas of grasslands that cannot be farmed. The grasslands are grazed, which means good hunting for surface finds in the short grasses. The eroded grasslands, referred to as badlands "breaks", are the best hunting areas because they expose the rocks and form large "beds" of rocks that await the hunter. The original Fairburn beds near Fairburn, South Dakota, is just such a place. There are many of these scattered from as far north as Rapid City and south into Nebraska. The Railroad Buttes between Rapid City and Scenic is an example.

South Dakota has a beautiful open quality to its landscape. At one time waysides on the interstate advertised that in South Dakota you can "feel free again". The openness and space do indeed convey that feeling.

Although lacking in trees, the grasslands, badlands, and prairie landscapes are very beautiful. State Highway 79 travels south from Rapid City for about 45 miles to Hot Springs and on the way passes by the community of Fairburn. The low sun on an early morning drive displays a beautiful pattern of shadows which highlight the eroding landscape of Fairburn country.

The locations covered in the text are primarily in southwestern South Dakota. However, some "spillover" occurs into Nebraska and Wyoming. The following map generally covers the geographic areas explored in this book.

The primary "original" locations well-known for Fairburn hunting are east of Fairburn, east of Oelrichs and south of Ardmore. The communities are located in South Dakota but good hunting areas are also located in Nebraska to the south of Ardmore. The general locations are badlands or grassland areas and are generally located by the arrows on the map.

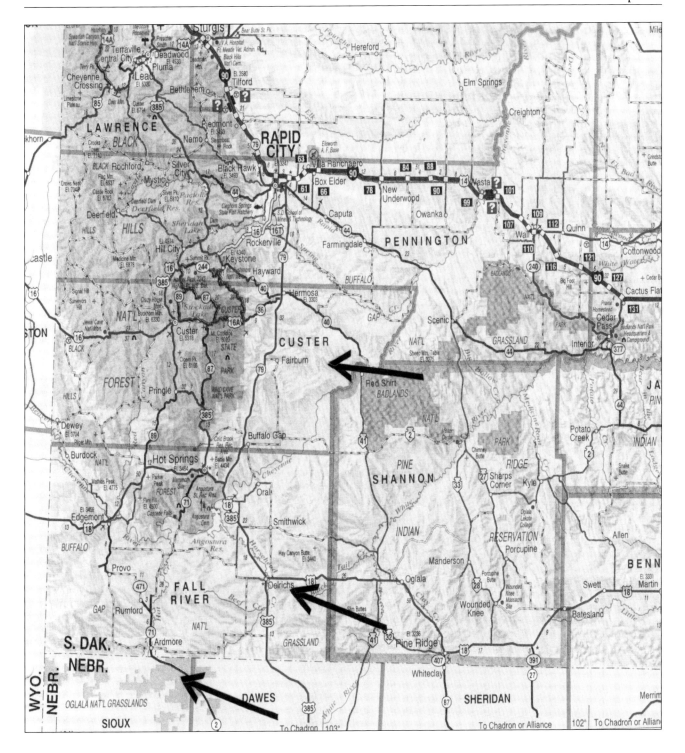

The agate collector or hunter who is not familiar with the wonderfully colored and patterned State Rock of South Dakota will find photos from several collections on the following nine pages.

From the Art and Ann Bruce Collection

This agate was referred to as "General Pink." Below is a section enlarged to show the spectacular detail. It was found at the Fairburn beds.

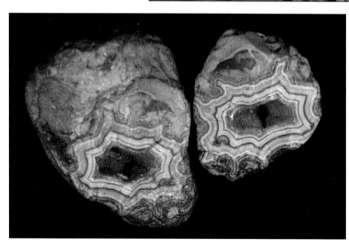

A pair of medium-sized agates that were found about one mile apart on different days. Art found one half and Ann the other. The agates were found in the "Oelrichs Beds."

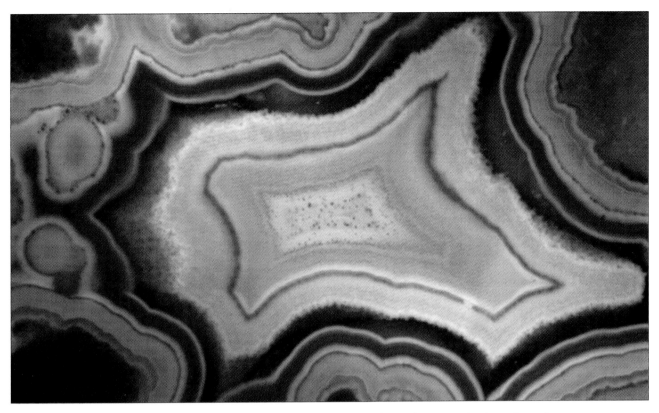

These close-up photos help to demonstrate the infinite variations in color and banding for which the agate is so well-known.

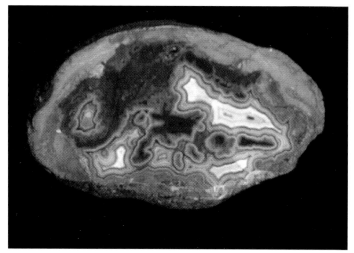

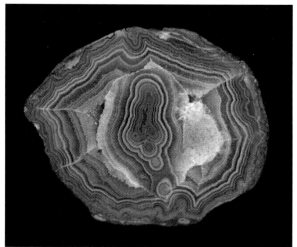

From the Jack Seger Collection

Jack acquired his collection, in part, from Mrs. Louis Miller after the death of her husband. Louis Miller once operated a rock shop in Hot Springs, South Dakota. The specific collection sites of these agates are unknown. Jack refers to the center photo as the "Queen" of his collection. The agates on this page (with the exception of photo on bottom right) are large agates.

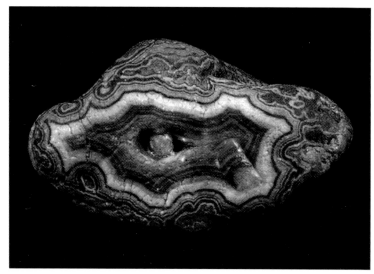

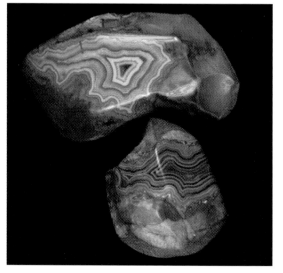

These Seger agates are small to medium-sized except for photo below. Large agates are seldom polished anymore because of their rarity. The majority of collectors enjoy them in the rough. This large agate would show tremendous color if polished.

From the George Burg Collection

George Burg is an "old time" collector. He has a collection not only of Fairburns; but many fossils, including cycads, fossil eggs, and other South Dakota materials.

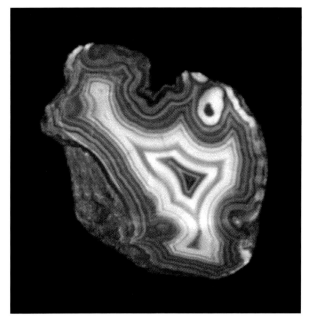

Collected near White River
north of Chadron.

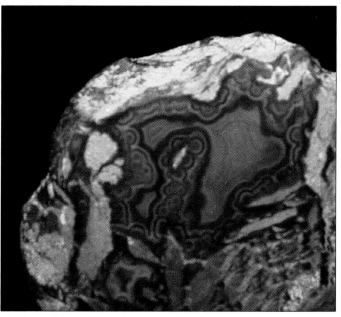

Collected in Nebraska grasslands. (large)

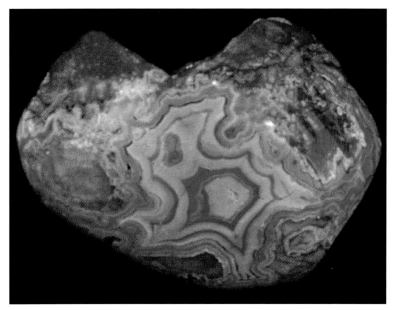

Collected near Lame Johnny Creek. (large)

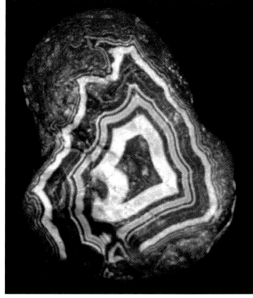

Collected at Nevis Draw near Scenic,
South Dakota. (large)

From the George Burg Collection

This pair of agates collected at Dow Ridge, (medium-sized) that represent the most colorful and intricately patterned agates found anywhere in the world. Along with the Dow Ridge agate (below). They make a spectacular display.

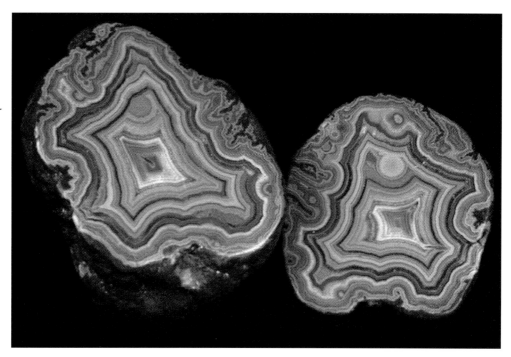

*Collected at Dow Ridge.
(east edge of the Black Hills)*

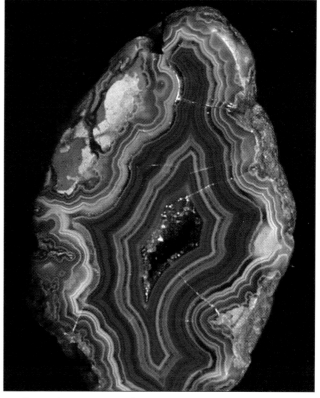

Collected near Pringle.

Authors Collection

Except for the agate pictured in the lower right corner, the agates on this page are from the author's collection.

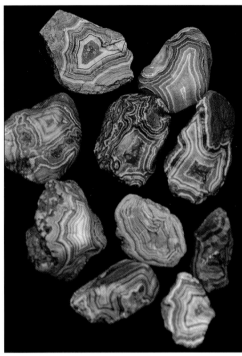

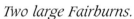

Two large Fairburns.

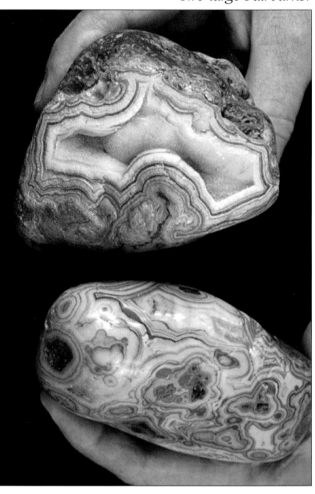

Small to medium agates are shown to demonstrate additional color variation and pattern. They have not been cut of polished.

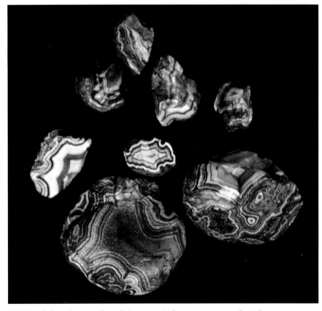

Agate found recently in Cheyenne River bed near Wasta.

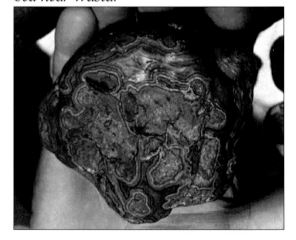

True black and white Fairburns are fairly rare. Many stones, once polished reveal red or yellow as their true color.

Author's Collection

Agates in the photo below and the lower left are all small, but colorful. They all have a polished face.

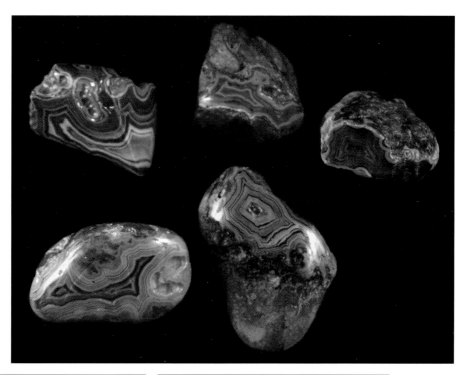

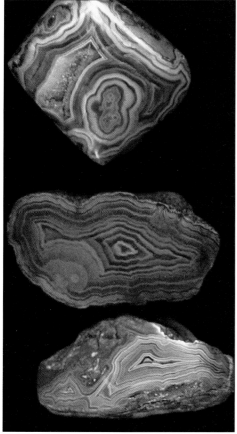

The small pairs to the left showed almost no sign of the beauty within before they were cut.

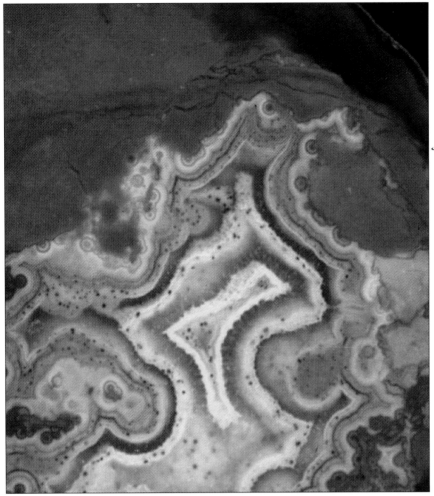

The close-up photo of this medium-sized agate demonstrates the pattern origination described in Chapter 8. Notice the small rounded formations where the pattern meets the matrix. As the banding forms it is layered over the rounded formations to form what is called the "holly leaf" pattern.

You cannot always trust the surface color. The chip off the lower left corner of the agate below reveals that the banding is not really black. Over the years...maybe centuries, the "weathering" and/or possible exposure to other chemicals and minerals, while buried in alluvial deposits, has changed the surface color.

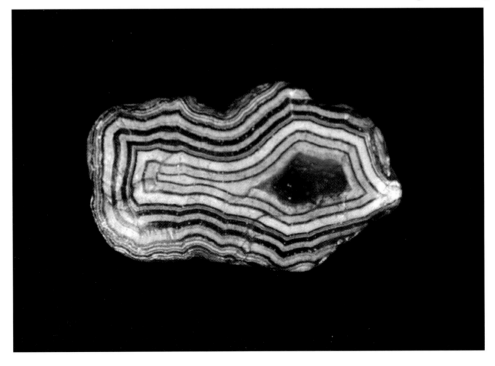

Chapter 2

Rockhounding: What Hooks a Rock Bass?

Someplace along the line a spark is kindled and a small flame of interest grows into a real love of rocks and minerals.

My grandmother's house had a large cluster of quartz crystals. There were arrowheads in an old cigar box that had been collected in the area. My great-uncle had a "rock garden" with water running through it. My mother told stories of finding small, bright Lake Superior agates in the gravels of the streams and gravel pits of the western part of Wisconsin. My great-uncle also had a beautiful red and white fortified Lake Superior agate ring. Those agates were also the topic of discussion on wonderful childhood trips to visit the Keweenaw Peninsula in upper Michigan where my aunt's family, on my mother's side, resided. Then there was a hometown minister who had a rock-collecting hobby where I could see some rocks cut. When I started college, geology was my initial interest. Employment however, looked dim, so teaching was the alternate choice. For a number of years, rocks were forgotten.

There was, however, enough interest that when my children were growing and I was looking for interesting ways to show them about discovering the world, we began tumbling some stones in a small tumbler. It got me back to a rock shop. It rekindled my interest. The renewed activity soon evolved into a family venture of rock hunting trips.

Beginning to Explore Rockhounding

The early rock hunting trips started with excursions close to home. We traveled to the Upper Peninsula of Michigan for Lake Superior agate and datolite. These trips eventually were extended to the Duluth/Superior area along the north shore of Lake Superior and the gravel pits west of Duluth, Minnesota. Along the north shore of Lake Superior, in the spring when it is still cold, we would hunt small Lake Superior agates washed in by winter storms. The rock exploring would continue until light dwindled and it was time for a driftwood fire on the beach as night approached.

The gravel pit deposits west of the Duluth/Superior area are excellent hunting locations for Lake Superior agate. The north shore of Lake Superior produces agate and (near Grand Marais), produces Thompsonite. We also ventured further north into Canada and visited the amethyst mines near Thunder Bay.

Discovering South Dakota's Prize Gemstone

About the same time, my wife, Mary Jane, and I had the opportunity to make our first excursion West on a combined business and pleasure trip. I had acquired a copy of *Midwest Gem Trails*. The cover had black and white pictures of intricately patterned fortification agates. It was the first time that I had been exposed to the large and wonderful rock hunting territory of South Dakota's grass-lands and badlands.

My trip to South Dakota included my first visit to Teepee Canyon to explore for Teepee Canyon agate like June Zeitner had described. She portrayed them as being similar to the fortified agate known as Fairburns, and painted an attractive picture in the Teepee Canyon portion of her book, *Midwest Gem Trails:*

> *Sometimes they will be solid jasper, sometimes they will have fanciful patterns of pinks, reds, yellows in measured concentric bands. If you are very lucky, you will find one like Chet Archer of Custer has, one of the most colorful agates in the world, a veritable Joseph's coat of the agate world.* —page 14

If you like agates with lines and you like color, Teepee Canyon and Fairburn agates are hard to resist after reading Zeitner's description. That book also described a comparison with the enhanced or dyed agates on the market. She painted a picture of that wonderful legacy of nature called the Fairburn agate. *Midwest Gem Trails:*

> *The true colors of a Fairburn can never be reproduced successfully by artificial means. A dyed agate may succeed in getting bands of pink alternating with bands of brown, but here are the colors, in order, of a Fairburn in front of me: brown, clear, pink, white, black, wine, pink, pen line band of wine, pale blue band, raspberry red—quite wide—white band set off by red, narrow black pen line band outlining the center which is yellow gold with a wine-colored triangle in the middle. Since these colors are produced by different means which often cancel one another, only nature can get them all together successfully.* —page 18

On that first occasion West, we had very little time to explore, but with a copy of *Midwest Gem Trails,* I began to follow the trail of the Fairburn agate. The visit planted the "seeds" which thereafter compelled me to explore and hunt those beautiful agates.

Rock shops in Wisconsin did not have Fairburn specimens. It was hard to know what was I really looking for. The badlands produces a lot of pretty (some banded) agate-type materials. I brought back some pretty "agates" from South Dakota. Although the local rock shop owner did not have any, the owner had seen them and assured me that my first finds were not Fairburns, prairie agates maybe, but not Fairburns.

Zeitner's book listed a number of people who had Fairburn agates, but none of them were connected with an address. One of them, however, was a rock shop owner, Bill Zieg from Alliance, Nebraska. I made the contact by mail and bought my first samples of Fairburns from him. Now I had a better understanding of the appearance of Fairburns. Zieg always tumbled his finds, so there was still the problem of what they looked like in the rough. Zieg was very helpful, not only providing me with the rocks, but sending me wonderful hand-drawn maps of hunting locations for various types of agates. An example of a map is shown on pages 17 and 18. The information on the map should still be valid as it relates to agate hunting areas.

On later travels, I stopped to visit Zieg and talk Fairburns. He told me, weather permitting, he

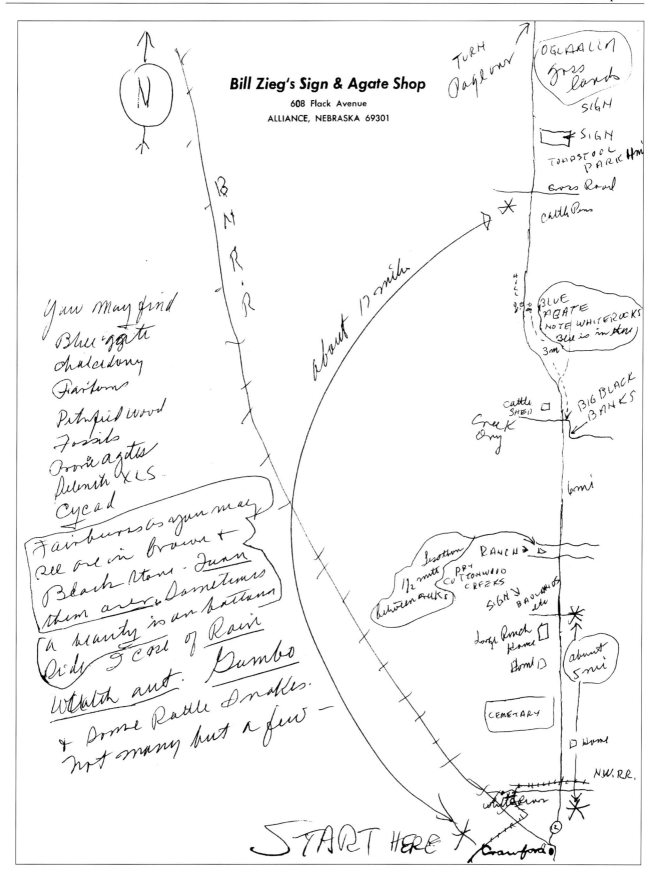

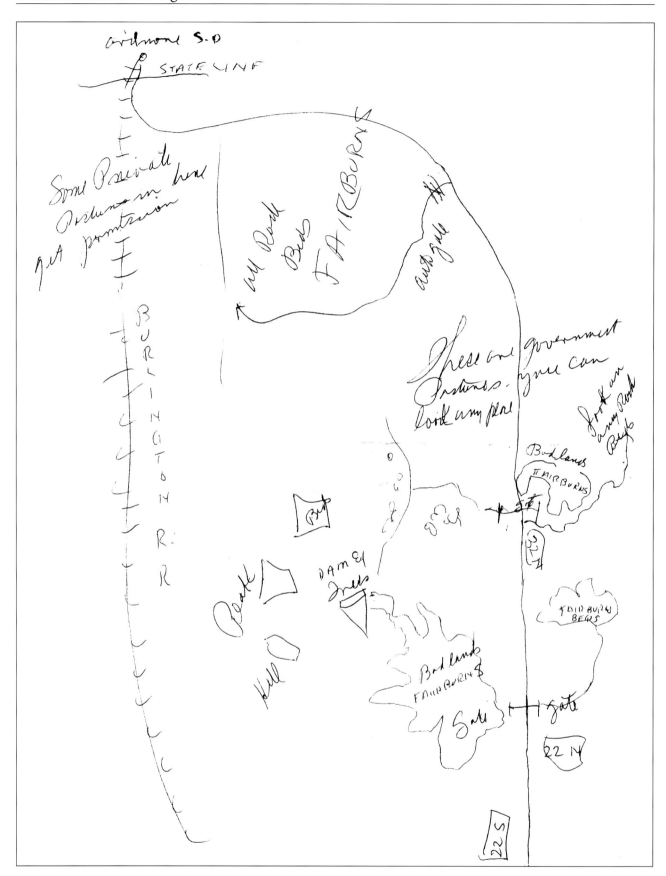

hunted every day. My understanding is that upon his passing, a substantial share of his Fairburn collection may have gone to the University of Nebraska.

The Fairburn Trail Continues; Scenic, South Dakota, and a "Famous Museum".

One of the next trips to South Dakota included following one of Zeitner's gem trails to look for black agate west of Scenic, just outside the west edge, of the "Big Badlands" (Badlands National Park).

It was my first time in Scenic, South Dakota. I planned to buy gas there and head out to hunt.

Towns are few and far between in South Dakota. Just because the name appears in fairly bold print on the map does not mean that there will be any services (or even people). There was no gas station. There was the Long Horn Saloon. "Scenic So. Dak. 1906" it said on the sign. I inquired about gas at the Longhorn.

No, there was no gas station, but the bar owner had elevated tanks in the back and we could buy fuel from him. That is, as long as we were willing to wait until he returned from a trip to sell a horse.

An hour and a half later, we were on our way.

Having no luck with the black agate, we stopped back in Scenic to enjoy the shade of one of the few main street trees while we ate our lunch. We were parked in front of a building surrounded by high woven-wire fence, next to the post office. While we were eating a pleasant-

Long Horn Saloon in Scenic, South Dakota.

looking elderly gentleman came out to the fence and asked the children if they would like to see a world-famous museum. The kids were naturally enthusiastic, and very curious. Mary Jane and I traded glances. What's this guy up to? What does he mean, world-famous museum in this desolate small town? There were no signs and the man came from a run-down building next to the run-down looking post office.

We were surprised, it was really quite a museum! And it was world famous because Clarence Jurisch kept a book and all his visitors would sign in. The museum had fossils. Our first immediate view was a Titanothere skull.

Jurisch also had agates, Fairburn agates. These are the ones I was hunting for!

"Where do you find them?"

Jurisch gave directions from Scenic:

- *Take Highway 44 west to Rapid City,*
- *You will see a gravel road just past the "five-line highline".*
- *Go south on that dirt road until you see rocks on the left with a trail that heads southeast toward the buttes and badlands breaks—you can see them from the road.*

These directions took me to the Railroad Buttes for the first time.

It was a beginning, but my station wagon was no match for that trail. Pickups and four-wheel drives have to be the order of the day to access good Fairburn hunting without a long walk.

Jurisch was one of the wonderful old characters of that area. Both Jurisch and the Longhorn Saloon were included in a 1981 feature article in *National Geographic*, Vol. 159, No.4, April 1981 on the Badlands. Clarence informed us that the Longhorn Saloon was the first building in town in 1906. As of the date of this publication, the Longhorn Saloon is still in operation and now there is a gas station. Jurisch has since passed away and I would assume his collection passed to relatives and other museums.

Although early hunting had produced no Fairburns, Zeitner's description of South Dakota's Fairburn agate and collecting areas continued to stimulate my desire to follow the trails. She accurately described "the Black Hills and badlands particularly, are so covered with interesting rock deposits and outcrops that hundreds of tourists each year are inspired to become collectors." The lure of the beautiful Fairburn agate and the multiple deposits of colorful rock had hooked me for good.

Clarence Jurisch at the door of this World-Famous Museum with son, Eric, and daughter, Julie.

Inspecting one corner of his fossil display.

Clarence proudly displayed his Titanothere skull.

Chapter 3

Learning about badlands and grasslands Fairburn Hunting

By this time, some Fairburns had been acquired by purchase and descriptions had been obtained on locations to hunt (including the famous rock beds near Fairburn). Still, I had made three or four hunting trips without ever actually finding a Fairburn. I was beginning to believe the casual conversations with people from the area that the Fairburns had all been picked up. There were vast areas covered with rocks, but I could not find one with a Fairburn pattern.

I did find agates. They had patterns and color. But I came to learn that they are called prairie agates or picture rock. If as a new agate hunter you find agates as shown in the top photo on page 22, they are not Fairburns. You can see, however, that they still have some beautiful color and banding.

It wasn't until I had the opportunity to track down Art and Ann Bruce, as a result of their auction in Hot Springs, that I began to really learn about present day South Dakota Fairburn hunting.

A new explorer to the area must first understand that most easy finds have already been picked up. The most likely locations are the newly eroded areas. The Bruces willingly went along with my wife and me to show us some of the better hunting areas to improve my chances of finding my first Fairburn. That first Fairburn came out of a grassland rock bed in what is known as the Nebraska beds. It was Art Bruce who also lead me to the location at the original Fairburn beds where I turned over the best Fairburn I have taken from that location. The Bruces told me that red and black rocks are the most promising. My first Fairburn, however, had a gray matrix.

Zieg had noted on his map "turn over the rocks". This did not really sink in until hunting with the Bruces. Hunting in the old Nebraska beds requires covering a lot of territory and turning over a lot of rocks. The hunter must get to know the general type of rock that needs to be turned. You can't expect the face of the Fairburn to be smiling up at you as they may have been in the past. The Nebraska beds have more grass and less active erosion. The new face up finds there have been few and far between.

At locations like the old Fairburn beds, where there is still a lot of active erosion, hunting the newly-eroded gullies and hillsides is the best bet. Erosion does continue to reveal new finds every year.

"Fairburns" Distinguished from Teepee Canyon or State Park

As I learned more about Fairburns, including Zeitner's descriptions, I became aware that there was a distinction that was drawn between the badlands Fairburns and rocks with similar fortification patterns that came from the Black Hills. State Park agate and Teepee Canyon agate were known examples of those fortification agates. Art and Ann Bruce refused to acknowledge any connection between the Teepee Canyon agates of the Black Hills and the badlands Fairburn.

Zeitner momentarily explored thoughts about the origin of the Fairburns in *Midwest Gem Trails*.

Prairie agates or picture rock. The slab at the back came from a large prairie agate found near Antelope Spring.

Do not expect a find showing a face like this. They are extremely rare. The photo is also included to show the "leakage/deformation" channels discussed in Chapters 7 and 8. Most writers now agree these channels are deformation channels caused by internal forces, forcing internal material out.

The photo shows the same pair as shown in the Bruce Collection on page 6. One-half is now face down to demonstrate to a new hunter that no pattern shows on the back side. Many Fairburns have this "generic rock" look. Regular hunters develop a sense of which rocks show promise so they do not have to turn over every rock.

With vast hunting areas virtually covered with exposed rocks you realize you cannot look at each one. Agate hunters eventually learn which rocks are more likely to contain an agate and turn those rocks.

She described the Fairburns as well as what she called Fairburn-type agate. The examples were State Park agate and Teepee Canyon agate. She noted that collectors theorized that there must be some relationship between these Fairburn-type and the real Fairburns. She posed the question: "Could the Fairburns have weathered out of a formation like Teepee agate or State Park agate?" That question became the center of interest for me. I kept asking the question, where were they formed? How did they come to rest in the Badlands? Were they formed in the Badlands? Theories ranged from formation in the Badlands to having been washed from the Rocky Mountains hundreds of miles away.

Zeitner described a bulldozed location near Wind Cave (Wind Cave National Park) where a Fairburn-type agate was found. She admitted that most of the characteristics of Fairburn agate were present, but then went on to list a host of reasons why she did not believe these were true Fairburns:

-Too much similarity.

-Bands (lines) too fine, too uniform.

-Too much gray matrix.

-Attached to a limestone layer.

-Coated (at times) with white lime.

She named them Fairhills agate because they were like Fairburns, but were from the Black Hills (not the badlands).

She further acknowledged that there were known agate outcrops that extend from near Buffalo Gap to the Wyoming border. According to Zeitner, "The best are between Pringle and Minnekata."

At the time, the conclusion sounded logical. Not only was she the expert, but all others that I had contact with voiced the same conclusion. Since no one that I knew at the time seemed to have any Fairhills agate, there was nothing upon which to judge.

Art Bruce continued to discourage and derail any thought of badlands Fairburn agates being related to agates found in the Black Hills. He despairingly referred to Black Hills agates...especially Teepee, as being "limestone agates." He assured me that no self-respecting South Dakota Fairburn collector would compare Fairburns to Teepee Canyons. With 40 years of agate hunting, a fabulous collection, and a lifetime of living in the Black Hills, who was I to take issue?

Teepee Canyon and State Park Inspire Further Inquiry

Still, it appeared to me that the exquisite fortifications and color of Teepee Canyon agate and State Park agate had too many similarities to Fairburns to simply dismiss them as "limestone" agates.

The curiosity persisted. My investigation included digging, mapping, and researching geologic history in search of the origin of the Fairburn. It also led me to inquire about the creation or genesis of banded fortified agates. It was obvious to me that some agates formed in volcanic-type rock because I had found Lake Superior agates in the basalt matrix in Minnesota and Upper Michigan. It seemed likely to me that the Fairburn agate had formed in some type of a sedimentary rock similar to the Teepee Canyon agate which is mined from limestone in the Black Hills. The Black Hills, however, did have its share of subsurface volcanic activity which confused the issue. That volcanic activity had pushed up the surface and created the granite peaks from which Mount Rushmore and Crazy Horse are carved.

Not far from these monuments, however, is Teepee Canyon, also known as Hells Canyon. In Chapter 7 of this book, you will find a description of the Currington Teepee Canyon agate mining activity. This mine, which is in limestone, later contributed geological knowledge to help follow the trail of the origin of the Fairburn.

In Chapter 4, I have related the thought processes and activities which I then believed would assist me in the search for the origin or source of the Fairburn. That search revolved around questions such as:

- Mining at the Currington Teepee Canyon Claim:
    How do agates form in limestone?
- Digging at the Fairburn agate beds:
    What layer do they come from?
- Hunting at the Fairburn agate beds:
    In what formations were they found?
- Topographic elevations at the Fairburn beds:
    How did it compare to Nebraska? Oelrichs?
    Could I find new areas by searching a similar topographic elevation?
- Geologic age of the agates:
    If I knew how old they were, could I trace them to the layers of rock of similar age?
- Chemical content:
    Could I make a relationship from the chemistry to somewhere in the Black Hills?

What further confused the issue is that Fairburns were found in the grasslands and badlands, along with fossils, cycad, petrified woods, chalcedony, and other agates such as prairie agates, bubble gum agate, and other miscellaneous forms of agatized material. What did all of this mixture in the Badlands mean?

John Paul Gries of the South Dakota School of Mines gave me a Black Hills summary which he used in part in *Roadside Geology of South Dakota* (1997). In Chapter 5 where I deal in general with the Black Hills geological history you will find part of the summary as written by Gries. You will readily see in that chapter why the Black Hills geology has contributed to the difficulty with determining the origin of the Fairburn.

Chapter 4

A Search for the Origin—Fairburn Territory Explored

The search for the place of an origin of Fairburns actually started because of the speculation about sources (Fairhills agates) in Zeitner's accounts in *Midwest Gem Trails.* Rumors of new Fairburn finds in the Black Hills and descriptions of thousands of acres of badlands and Black Hills hunting inspired me to find a way to scientifically locate likely new Fairburn areas.

When one visits the Badlands again and again searching for rock, it occurs to you that the agates that are exposed on the surface at Fairburn and Oelrichs must be associated with certain formations. These formations would undoubtedly contain geologically similar alluvial gravels.

Origin Theory Number One—Stratigraphy

If I could identify the layer of badland formations where Fairburns were found and locate related stratigraphic layers, I should then be able to identify locations where Fairburns would be exposed. It seemed logical that layers would be exposed at similar elevations throughout the grasslands and badlands. A review of the topographic maps would allow me to locate eroding grasslands area of the same elevation and thereby find locations where the Fairburns should be eroding to the surface.

In addition to locating more Fairburn beds, stratigraphy should also allow me to relate the layer to geologic events in the Black Hills. That, in turn, would help lead to the rock unit from which they were eroded and the origin of these agates.

There are few (very few) Fairburns yet discoverable at Fairburn. After several attempts, however, I did find several Fairburns (one about the size of a tennis ball) in place as they were eroding out of the gravel layers.

I also excavated and screened a gravel "lens" to attempt to dig a Fairburn. Gravel lenses form from stream action that leaves gravel deposits layered in the shape that looks like a camera lens viewed from the edge or side. After searching pails of screened stones of various sizes, I was not able to find even a "chip" of a Fairburn. The material from the gravel lens was identical to the material in which I had previously found the Fairburns. As a result of having previously found Fairburns in those similar gravels, it seemed logical that the sand/gravel lenses are where the Fairburns have been deposited from the eroding Black Hills. The erosion of those sand/gravel lenses are why new stones are continually being exposed and collected.

Once I had this information, I thought that I could correlate it with the Nebraska badlands breaks and the Oelrich beds which lie between Nebraska and Fairburn. No such luck! The stratigraphic units do not follow the same topography or elevations.

The geologic formations of the lands east of the Black Hills **do not** order themselves in neat, level, and even layers. My search of the geologic literature lead me to publications sponsored by the South Dakota School of Mines. One needs only to follow the geology field trips in the School of Mines publications to discover that the layering of the Badlands is a hodge-podge. Not only is it uneven and nonconforming, but also the layers have been so blurred as to identification they are, at times, hard to identify once located. It's a real geological mess.

As an example, when describing terrace gravels southeast of Rapid City, in the *Geology of the Black Hills, South Dakota and Wyoming,* Philip R. Bjork and J. R. McDonald included this passage:

> *View of Railroad Buttes from the Farmingdale Terrace of Plumley (1948) the lowest of surfaces along Rapid Creek. The age of these terraces has not been established and remains a major stratigraphic problem which has a major bearing on the physiographic develop-ment of the Black Hills.   —page 211*

The publication is a field guide book and spends many pages trying to reconcile the numerous descriptions of the same formations that have been given historically. The *Guide to Cenozoic Deposits,* also demonstrates the complicated erosional patterns that first created and built the Badlands and then subsequently eroded it away again. The result of these uplifting and erosional processes resulted in a situation where (as least to the amateur) there is no relationship right side up or upside down or otherwise between the layers in the Black Hills and the layers in the Badlands.

There is still disharmony among the writers which is pointed out in the *Guidebook to the Major Cenozoic Deposits.* The writer notes that some geologists believe that the Black Hills was eroded to its current level by 32 million years ago. He poses the question, if that is so, then where did all of the erosional deposits in the Badlands come from after 32 million years ago?

For example, where did the Sharp's formation described in the *Guidebook to Cenozoic Deposits* obtain its alluvial deposits if the Black Hills had already been eroded away?

There are several publications and maps that demonstrate that ancient history has eroded the Black Hills uplift in a primarily eastern direction. The *Guidebook to the Major Cenozoic Deposits* actually has charts to identify the names which have been used to describe the same or similar formations. It also relates that geologists from the early 1900s through the present have struggled to define the geologic formations of the area. It was clear that this area of inquiry had led to a dead-end.

For the readers information I have included a diagram published by the American Geological Institute in the *Geology of the Black Hills, South Dakota and Wyoming (2nd Edition)* which portrays a graphic illustration of the stream flows of the Black Hills that resulted from the uplift.

As an amateur, it suffices to say that if you really want to learn about the multitude of forma-tions, such information is available. There are chronologies, charts, maps, tours, and photographs that can be studied, studied, and studied. The Geology Resource Center of the School of Mines has a wealth of information for a small price. Don't miss the Museum of Geology on the campus. It is a wonderful museum with great dinosaur fossil reconstruction as well as a large rock display. They display some beautiful agates including Fairburns and Teepee Canyons.

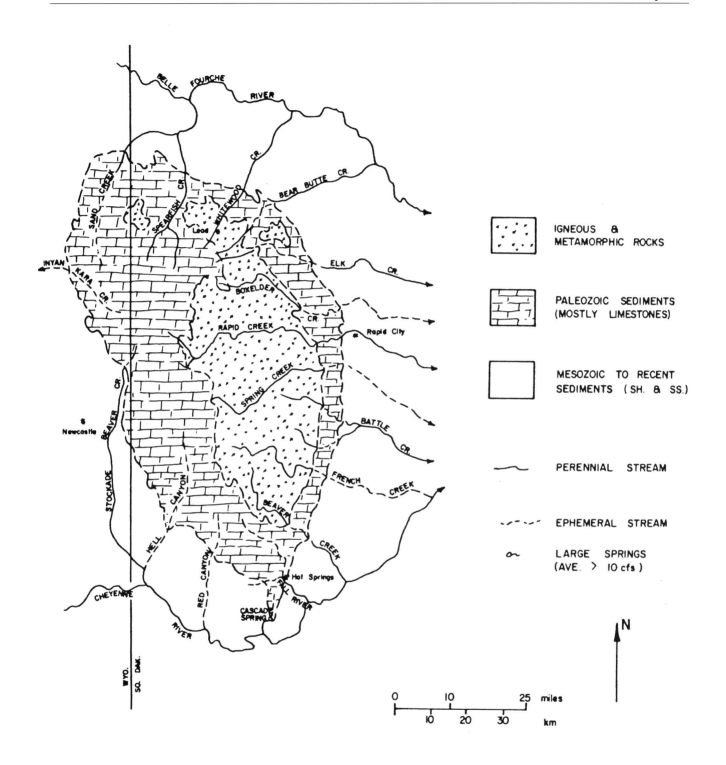

*Map of stream flows arising in and around the Black Hills
demonstrating the predominant eastward flow of those streams.*

The following figures taken from *The Guidebook to Major Cenozoic Deposits* demonstrates the continuing erosional directions as related to recent ice ages. The glaciers themselves never covered the Black Hills.

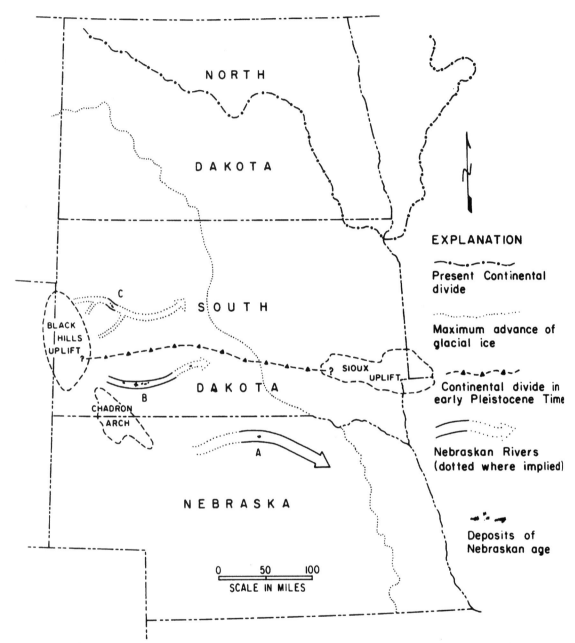

Figure 9. Outline map of North Dakota, South Dakota, and Nebraska showing the relations between uplifted areas, present and paleo continental divides, the maximum advance of glacial ice, and three stream deposits of Nebraskan age. A. The Sand Draw locality. B. The Medicine Root gravel. C. Unnamed gravel mapped by Pettyjohn (MS).

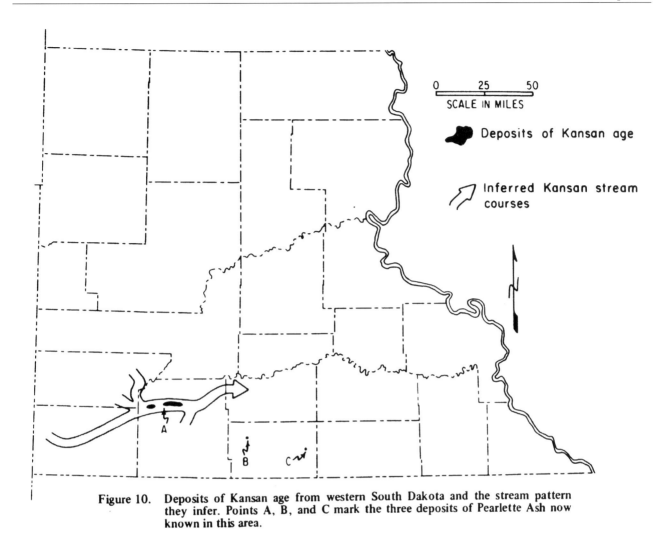

Figure 10. Deposits of Kansan age from western South Dakota and the stream pattern they infer. Points A, B, and C mark the three deposits of Pearlette Ash now known in this area.

Origin Theory Number Two: Chronological Dating of Fairburns

How old are the Fairburns? Having failed to be able to make any "scientific" relationship of the stratigraphic units to topography, it occurred to me, if rocks could be dated as to their age of origin, then I could relate them to a geologic unit of the Black Hills.

This turned out to be a very short search. The dating of the rocks, I am informed, does not work when the rocks have been exposed to surface elements. If there was a Black Hills origin, then the Fairburns had been exposed in the Black Hills, washed into the Badlands and buried again, only to be later washed into the open again where they were being found at Fairburn. Fairburns I took to the School of Mines could not accurately be dated by traditional dating methods.

How else might someone research the background of these Fairburns? Since I was at the School of Mines and these students were learning about geology, in particular this area of South Dakota, what better source than to find a student who would be interested in a term paper or thesis on the subject of the origin of these unique Fairburn agates?

At that time, I was visiting with a geologist and instructor at the School of Mines because I was hungry for information. This gentleman thought it would be an excellent idea to post the topic for interested students who may have wanted a subject for a research paper. Not one student even inquired! Months later when it was clear that no students had any interest, I asked for an explanation. The instructor could only relate that the lack of interest was due to the fact that:

1. Geologists need jobs following graduation;
2. When research is related to gold, oil, or other precious commodities, you would have a much better chance to obtain employment.

It was obvious to these students that Fairburns did not translate into money or jobs.

Origin Theory Number Three: What About Chemical Composition?

If you know the chemical composition of the Fairburns, would you not be able to relay that chemical composition to Teepee Canyons, State Park, and those Fairhills agates that Zeitner talked about?

Armed with specimens of these agates and optimism, I met with Gries who had also agreed to spend time with me on the subject, and who, himself, had some interest in Fairburns.

In our discussions, I became aware that he recalled the location of a seam of Fairburn-type agate in a canyon known as Redbird Canyon on the western edge of the Black Hills. He had come across this seam of agate in the canyon some years before and had described the location in a paper dealing with the geology of the area. Also found, however, on his return that as soon as the paper had been published, the agate had disappeared!

He responded to my new theory of how to find a Fairburn source with a kind smile. Unfortunately, the makeup of agates is almost entirely quartz. The composition of quartz is usually near 100 percent SiO_2. There are only a few trace elements that color the agate. They are the very common elements (such as iron) that are found universally and therefore, chemical composition would be of little or no use.

Silicon dioxide (SiO_2) occurs as crystalline quartz, cryptocrystalline, chalcedony and amorphous opal. It is the primary ingredient of sand and chert. Such common ingredients did not allow for geographic association to any specific locality.

Subsequent studies of quartz and its inclusions has confirmed for me what he already knew. Chemical composition was nothing more than another dead-end.

It seemed at this point that being able to establish any logical origin by direct geologic information was not possible. The Teepee Canyon agate was the only Black Hills agate that could be found at or near its point of origin. The Teepee Canyons seemingly distinct appearance seemed to set it apart from the Badlands Fairburns.

As Zeitner had stated, the Fairhills agates and State Park agates, while possibly related, seemed to be distinguishable from Fairburns.

Chapter 5

Where Else or What Else is There to Look For?
Is There a Black Hills Origin?

At this point, I had practically exhausted all ideas I had to determine the origin of these beautiful agates. The fact had not, however, discouraged my hunting and I continued my spring and fall visits to South Dakota.

To a rock hound, weather is always a factor. This is especially true in South Dakota because of the temperature extremes and the terrible badlands gumbo clays. South Dakota's temperatures on average are warmer than Wisconsin even though they are near the same latitude. South Dakota in April and May (while it can be wet) is very comfortable for agate hunting. The same holds true in the late part of September and through the month of October. If the winter is open (no snow), rock hunting can be done even in December and January.

Agate hunting over the summer months can be brutally hot in the badlands where the temperatures often exceed 100° F. The risk in the spring is rain or snow which on the prairies can move quickly and turn the gumbo clays into a virtual tarpit. In the spring, however, there is an added bonus. The air is constantly filled with the beautiful song of the western meadowlark. Dozens of them will be trying to out-sing each other as the sun warms the spring landscape. One other benefit of the spring and fall is that you are visiting in the off-season. Motel rooms are plentiful and about one-half the price charged during the summer tourist season.

Information from rock shops in Rapid City, Custer, and Hot Springs provided me with one more interesting hunting option, the gravel pits along the Cheyenne River. The Cheyenne River on it's course circles around the south end of the Black Hills and then crosses the Badlands on its way north. It lies between the Badlands National Park and the Black Hills.

The Cheyenne actually cuts diagonally across the erosional deposits that washed eastward from the Black Hills. The gravel deposits of the Cheyenne are filled with all kinds of good things for rock hounds. Fossils, such as marine fossils and dryland fossils including horse and camel teeth, various types of petrified woods, and all the various types of agates of the badlands can be found there. The agates include Fairburns, prairie agates, bubble gum agates, and other agates with various local names.

A word of caution at this point, the gravel pits are all privately owned. Unless you have specific permission to hunt, you are trespassing. The gravel pit operators, while usually cooperative, have been increasingly cautious over the last few years.

There are areas of the Cheyenne River that pass through the national grasslands. These areas are open for hunting in the river gravels or in the gravels eroding from the banks into the river. Be especially careful along the river. There are still rattlesnakes in South Dakota!

The fact that the Cheyenne River accumulated agates was no real surprise. Neither did it provide me with any new clues because it was cutting across and through many levels of badland deposits.

The jumble of rock hound materials in the gravels led me back again to the question of the genesis or formation of agates. If I just knew how the agates were formed. That would probably be helpful to figure out whether these agates formed along with the petrified woods, cycads, fossils, and other agates. Would that not help in an attempt to find where they came from?

Would it be significant if the petrified woods and cycads, as an example, were found in the same eroding formations as the Fairburns? Would they not have lived, died, and been petrified at a time when the formation of Fairburn agates was occurring? Because I could not relate Fairburns to any formation, this also turned out to be useless speculation. In addition, cycads, for example, came upon the scene 225 million years ago and were prominent through the time of the Black Hills uplift 70 million years ago. The prominent Oligocene fossil beds of the badlands are estimated to have been deposited between 25 to 36 million years ago. As all of these layers weather out they are difficult for amateurs to decipher.

When the grasslands and badlands were wet, or when the weather was poor, the gravel pits provided a very interesting variation to hunting. It offered hunting for agates in piles of rock that was freshly excavated from the Cheyenne River Valley. I have found Fairburns in the gravel pit at Oral and also in the gravel pit at Creston. I had been told Fairburns have been found in the gravel pits at Wasta. It was a recent privilege to view and photograph a nice Fairburn find which came from a gravel bar in the Cheyenne River near Wasta. That agate can be seen on page 12.

All the rock shops have a special display case showing Fairburns. There was usually a new color or pattern of Fairburn which is not represented in my collection so I would make occasional purchases. These Fairburns would come from local collectors and rock hounds. They are purchased by rock shop owners for resale. Regular purchases (as much as I could afford) were also made from the Bruce's, who were reducing their collection anticipating a move to a smaller home.

An Unexpected Discovery

My business occasionally had me flying West. On one occasion, I was able to arrange a stop for a couple of days for a little rock hunting, visiting friends, and rock shop stops.

On this occasion, after looking at and discussing a Fairburn purchase, I asked the rock shop owner if anybody had been bringing in any other local material. The owner promptly said yes, he bought some good material several months before. He was going to polish some of the material and display it. The material had come from the Black Hills and was identified by location. Because he wanted to show the local rock hounds that excellent fortification agate could still be found in the Black Hills. It was his intention to create a display which would be shown at one of the upcoming rock shows.

The material (approximately 40 to 50 pounds) was in the back room in cardboard boxes. A couple of pieces had been cut, but most of it was just rough as it was found.

The material was beautiful! It did not appear to be State Park or Teepee Canyon material. It was colorful, fortified, and in my opinion, it was Fairburn material!

I asked where it came from.

"Antelope Springs, S & G Canyon, and near Pringle," was the response.

"Where are these places?" I asked.

The shop owner was only able to give me general descriptions, but there was no question that they had come from the Black Hills and not the grasslands or badlands.

"Would you tell me who you bought them from?"

"Sure, "Billy," he works for a local car dealership. I can't remember his last name, but if I make a phone call or two, I think I can come up with his name."

After one call, the shop owner came back and gave me the name of Alfred "Bill" Stein.

The material was tremendous as it relates to pattern and color. I had to have that material! This was the first real additional information to identify a source that had surfaced. The owner wanted a pretty good price, but price was not the object. I wanted that material and I wanted to visit the locations.

The original purchase from the rock shop was shipped home because of the weight. Each box had various agates of patterned and colored Fairburn-type agates. They each had some distinctive coloration that seemed to be limited to that locality. With regard to each group, however, it seemed that there already was in my own collection a Fairburn with that same type of coloration or banding. It was interesting that several Fairburns that I had previously purchased also had rough or jagged exteriors to the matrix. The Fairburn agates, other than Teepee Canyon, that came from Art Bruce had always shown the smooth, "tumbled" exterior of the Fairburns. This was the signature of the alluvial deposits in the Badlands. If a Fairburn would have left its original home in the Black Hills and then been tumbled into the Cenozoic formations of the Badlands millions of years ago, it would have been smoothed by stream action on its journey to the Badlands.

The agates that came from Antelope Spring, S & G Canyon, and Pringle and did not appear to have been tumbled or smoothed. Many of the pieces were badly fractured. The material was obviously Fairburn, but the matrix to the material was jagged and rough, demonstrating that it had not been subject to alluvial forces. Were these agates at or near their source or place of origin?

After having the material shipped home, I began to look back on the story told in *Midwest Gem Trails.* It sounded just like the material from the area of Wind Cave, that had been labeled Fairhills agate.

Because of the condition of the material and the limestone deposits, Zeitner had distinguished it from the Fairburn agate of the Badlands. This same type of condition, however, told me a story that fit perfectly with the limestone layer origin for Fairburns.

First, the limestone matrix is agatized into jasper or chert where it is adjacent to the fortification pattern. When the exterior is tumbled away, there remains a hard matrix around the agate. This is common to all hills agates whether found at Teepee Canyon, Antelope Springs, S&G Canyon, or Pringle. It is also common to grassland and badland Fairburns.

Second, the Fairburns which were tumbled into the Badlands were exposed for a period of time to weathering. Eventually, however, they were covered by additional erosion and ash deposits and buried deep in the badlands alluvial deposits. In recent geologic history, they have again been exposed in the grassland and badlands gravels as they are eroded out again.

The agates remaining in the Black Hills deposits, on the other hand, had been exposed through erosion at the surface, but then had remained exposed. They have been subject to weathering for millions of years longer than those same agates that were tumbled into and buried in the badlands. It would be expected, therefore, that Black Hills agates are subject to many more weathering fractures.

The Teepee Canyon agate is a further demonstration of the difference. Teepee Canyon, which has been mined, has never been directly exposed to the elements. This agate is, therefore, essentially free of fractures other than those that come from the miner's hammer.

Not only agates, but many of the rocks in the Black Hills have white superficial deposits of limestone. These deposits simply formed at a later date. The agate that lay either exposed or near the surface in the Black Hills simply accumulated some additional limestone or other mineral deposit. These deposits on the exterior would have occurred after the Black Hills erosion slowed about 30 million years ago.

Renewed Black Hills Exploration

These new locations had to be visited. Were these sites similar to the Teepee Canyon area which was a well-known Black Hills agate producer?

Bill (Alfred Stein) was hesitant at first, and rightfully so. Rock hounds are usually suspicious of someone else finding their "best fishing hole", especially when they come from "out East". We soon got past the suspicious stage and he agreed to show me these new locations.

When asked how these sites became known to him, he related that he had worked for the U. S. Forest Service which manages the Black Hills National Forest. He was one of the men who would trim and cut underbrush in the forest. This is routine work in the Black Hills along with other forest management needs.

On one occasion while working in the Antelope Spring area, he noticed a fortification agate in the roots of a windfall. Bill's father had been a rock hound and Bill knew a Fairburn-type agate when he saw it. He mentally marked the spot so he could return to hunt later. He still has not located the spot, but while searching for it he discovered the Antelope Spring deposit. He has been back numerous times (several with me), but the original location is still unexcavated.

As a historical note, Art Bruce had also suggested another method of hunting in the Black Hills. Look in the roots of downed trees. The pines that cover the hills have a shallow root system because they spread out in the thin soil over the top of the rock. When a tree is downed by a windstorm, the roots are then exposed. They carry rocks up with them exposing these once-buried rocks to the elements. They are washed off by rains and are held up in the air as they remain entangled in the tree roots. Art had related how he and Ann had found some good Teepee Canyon agates in such a manner after big storms.

When in the Black Hills, I noticed that such an event could indeed expose the rocks that had been buried below the soil. It was an interesting coincidence that the discovery of the Antelope Spring location arose from the observation of the roots of fallen trees.

On one of our hunting trips near Antelope Spring, we also found a field with a nice deposit of "prairie agates" on the surface. Since then, there have been several locations excavated containing banded or patterned agates. Had these agates been tumbled into the Badlands, they would have been called prairie agates or picture rock.

Field Visits to the Agate Sites

My very next trip to South Dakota included a full day tour of Black Hills locations. Enough time was spent at each location to find at least some evidence of fortification agate to confirm, that indeed, this is where the agates have been found.

Antelope Spring is in the northwestern part of the Black Hills near Wyoming. The other two locations S & G Canyon and Pringle are on the western and southern areas of the Black Hills. Now I was armed with agate location information which could be correlated to Black Hills geology.

With the help of the maps purchased from the School of Mines, an identification was made of each location where agate material had been found. Each of the locations correlated with the remaining limestone border that ringed the Black Hills. That border can be seen on the Geologic map of the Black Hills which can be found on pages 36 and 37.

Gries also identified the location where he found the Fairburn-type agate in Redbird Canyon (southwest Black Hills). The limestone border where these agates were being found was known as the Minnelusa formation. Gries traveled with me to show the location of Redbird Canyon. We also visited the Currington Teepee Canyon mine. He could identify the formation from which the Teepee Canyons were being mined was the Minnelusa. Each other location can be identified as being in or near a remnant of the Minnelusa formation that continues to encircle the Badlands uplift. When looking at the photography section containing the comparison photographs (pages 43 to 52), you will see a nice Fairburn specimen found by George Burg in Red Canyon north of Edgemont, South Dakota. That location is also in the southwest Black Hills.

All Trails Point to the Same Formation

It can be seen from the Geologic map of the Black Hills and the diagram on pages 36 and 37, that the Black Hills uplift caused the erosion of a huge area of the Minnelusa formation. It is my belief that the eroded Minnelusa contained the badlands Fairburns that we now find in the grasslands and badlands. The erosion stopped before it took all the agates from locations such as State Park, Teepee Canyon, Antelope Springs, S & G Canyon,and the Pringle or "Fair Hills" location between Pringle and Minnekata.

Dow Ridge, not mentioned before, is located at the southeast edge of the Black Hills near Custer State Park. It is private land. George Burg, who has permission to hunt, has a couple of the most beautiful specimens found in that location in the eastern Black Hills.

There is one other location that has been explored on private property near Minnekata. At this point, it appears that the location is an "interior" deposit of gravel. This is an area where materials tumbled toward the badlands area, but never really left the Black Hills. Ann Bruce, however, assures me that they have found Fairburns at this location north of Minnekata. Just before the publication of this book I was also shown a 10-pound Fairburn from the Minnekata area. Burg has now shown me where Zeitner's Fairhills agates were found, which is not far from Pringle in the southern Black Hills.

The Minnelusa Formation still has substantial presence in the Black Hills. It is exposed at the surface completely encircling the Hills.

Now knowing the finds that have been made, it is reasonable to conclude that there are many as

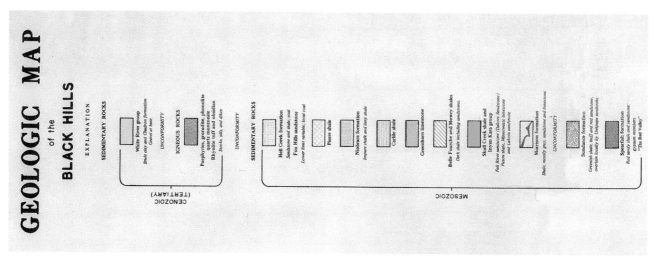

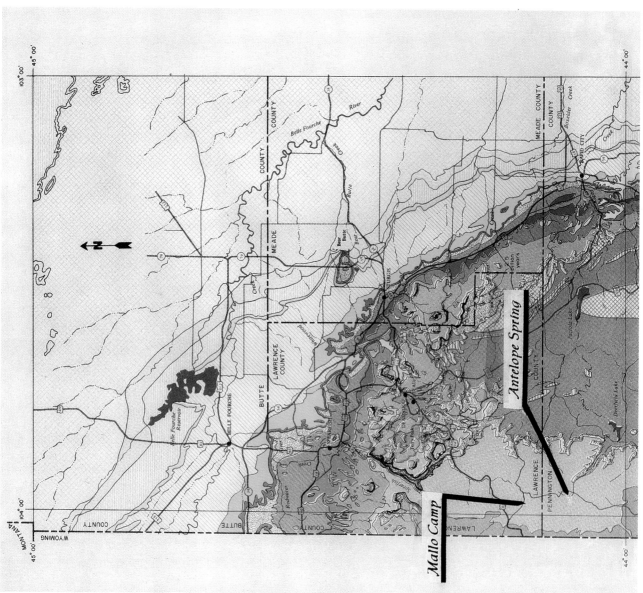

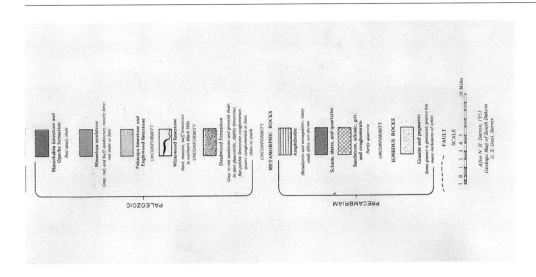

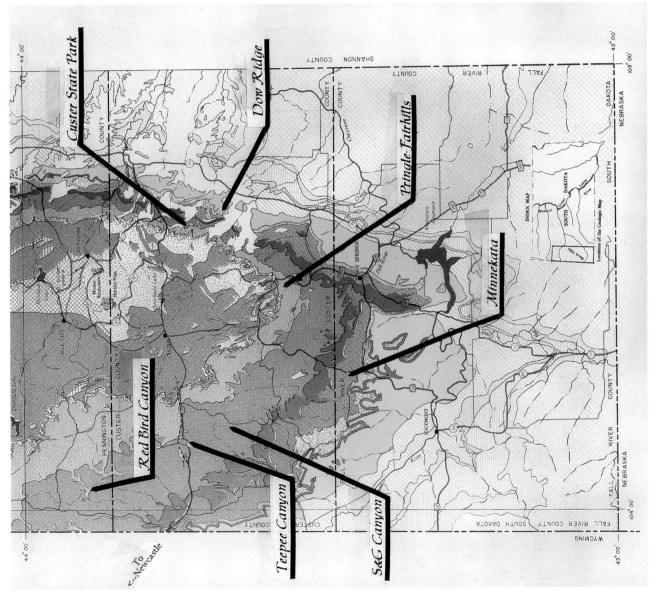

yet unknown Fairburn locations in the Black Hills. Bill Stein has shown some of the Antelope Spring agate to an old-time collector who has identified another location called Mallo, near the Wyoming border, where a fortification agate of a similar nature was found. The old rock hunter referred to it as "mallo agate".

How does one go about finding additional locations? The only answer is to search those areas where there are remnants of the Minnelusa formation or where the Minnelusa formation is the dominant rock formation at, or near, the surface.

Jack Seger (just before this book was published) supplied some information from the Internet on Fairburns. The entry asserted that a Fairburn type agate had also been found in place in a limestone formation near New Castle, Wyoming. This find would be consistent with the widespread formation of pockets of Fairburn agate-bearing limestone throughout the Black Hills area.

The source and origin of this most beautiful and colorful agate is the massive and varied Minnelusa formation of the Black Hills. (My apologies to Ann Bruce for relating "limestone" agates to Fairburns.)

Some argument can be made that at least some of these agates formed in the limestone below the Minnelusa layer known as Pahasapa (Madison) Limestone. The Pahasapa (Sioux Indian for Black Hills) is also a massive layer extensively eroded in the same manner as the Minnelusa.

My conclusion is based upon the fact that at least three locations in the hills, agates are found in place in the Minnelusa layer. The other locations are closely associated with remnants of the Minnelusa.

Why is the Fairburn So Unique and Varied?

Why do these agates come in such a variety of colors and patterns? What gives it such a variety? The Fairburns were formed in numerous pockets of zones within the Minnelusa formation. That likely would have occurred at or before the Black Hills uplift began about 70 million years ago. That uplift caused the Black Hills to erode into the badlands. Approximately 60 to 70 million years ago, the Black Hills uplift would have begun to expose the Minnelusa formation. The relationship of the Minnelusa Geologic time period can be found on pages 54 and 55.

The Minnelusa formation in the diagram below is represented by one of the sedimentary layers shown in the cross section. You can see from the diagram that the Minnelusa layers would have been tumbled together with layers from both above and below. Also, because the agate formations were individual as to specific locations (such as Teepee Canyon or S&G Canyon), each of the materials from each location was blended together as the materials eroded and moved to the east.

As the uplift begins, stream erosion begins to carry gravels from the Black Hills eastward. See also the stream flows on page 27.

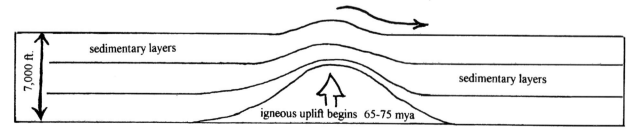

The continuing uplift causes simultaneous erosion of deeper layers thereby "tumbling" the layers together.

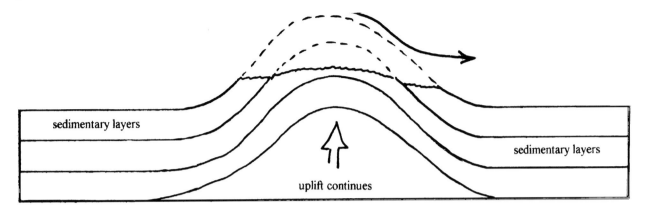

The gravels (so-called "Terrace Gravels") of the badlands contain cross sections of over 350 million years of sedimentary rock, including the Minnelusa formation. An estimated 7,000 feet of rock was eroded primarily to the east.

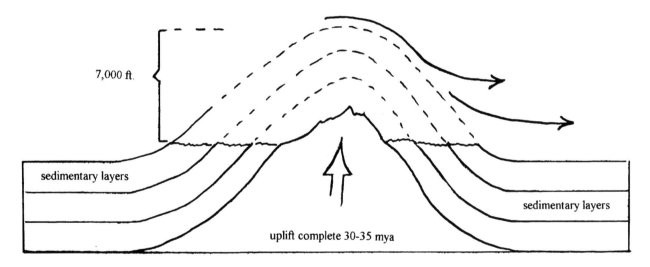

The Fairburns are fairly similar in the sense that they were all fortification agate. The colors, however, vary greatly from location to location. As noted, the Minnelusa was not exposed all at one time. It was gradually lifted at the center and then (over 30 million years) eroded at the edge of the igneous uplift. These multiple agate-bearing gravels were mixed and tumbled and eroded into the lowlands east of the uplift throughout 30 million years as what would eventually be badlands formations received deposition from the Black Hills. Agates may be scattered throughout any formation that would have developed by receiving erosional or alluvial deposits from the Black Hills.

In today's world, therefore, we can identify a certain coloration as coming from Teepee Canyon, or State Park, or Pringle, because we can find agate near the source. There were likely, however, thousands of other locations (long ago eroded) that were all jumbled together in the badlands. Each location made its unique contribution to the wonderful variety of colors and patterns. Although originating in many different areas they were tumbled together and are known as the fortification agate identified as Fairburn.

An argument to distinguish Fairburns not only from the "hills" agates but also from various badland locations has been the color of the matrix. By matrix, it is meant that the host rock which surrounds the pattern. Some agates are tumbled free of matrix, but most Fairburns have some remaining.

The same question had been explored in *Midwest Gem Trails:*

> *One night at Archers we tried a test, laying out pieces of South Dakota agates from many localities and trying to tell what distinguished each. Really, slabs of Nebraska agates with their brown matrix and Teepee with its brown matrix were all but identical. And many of the others could have been from one place as well as the next.* —*page 15*

The explanation for the similarity is in large part what this book is about. The explanation for the variable matrix colors is likely the Minnelusa formation itself. The formation is massive and variable within itself. *The Guide to the Stratigraphy of South Dakota* (page 123) contains descriptions of the Minnelusa by eight different writers. The colors used are white sandstone, mainly buff, and red to brick red. It is said to range from 75 feet to 1,300 feet in thickness. One writer describes three distinct layers within the Minnelusa. "(1) Minnelusa saccharoidal sandstone, 200 feet; (2) Minnelusa (alternate series), 300 feet; (3) Minnelusa white sandstone, 100 feet."

No publication has yet made reference to the origin of Fairburns, other than *The Roadside Geology of South Dakota.* Gries verified what his thoughts had been and my search had concluded.

Fairburn Agates

> *Terrace gravels associated with French Creek below Fairburn contained agates that were probably eroded from cherty zones within the Minnelusa formation farther upstream.*
> —*page 322*

As I have demonstrated, however, Fairburn agates are far more widespread than just the old Fairburn beds of French Creek. Not only are they widespread in the grasslands and badlands, but throughout the Black Hills as well. In a personal contact, Gries advises that it was purely oversight not to have included a far larger area.

In the photos on page 43, you will see what look to be agates from Teepee Canyon. Only one of the agates was found at Teepee Canyon. One of the agates came from near Pringle. The other two came from the badlands gravels near Oral.

The badlands gravels, therefore, produce agates that look like the traditional Teepee Canyons. George Burg also found this coloration near Oral. Another Hot Springs collector also found a wonderful specimen that was a Teepee Canyon look-alike.

Burg believes that they were washed out of the Black Hills near Teepee Canyon and washed down the smaller creeks to the Cheyenne River. The Cheyenne then carried them to the badlands near Oral. The distance of that stream course is about 75 miles. Assuming Teepee Canyon agates can travel such a route...then hundreds of other deposits could do the same thing as the Black Hills eroded. Each original deposit had variations in color and pattern. Therein lies the Fairburns' fantastic variety.

Burg also believes that it is the orange in the color that is the most scarce. The Antelope agate in my collection has a lot of orange coloration. The Antelope agate location is also the most distant from the badlands and the least likely to be carried to the badlands. However, the badlands and grasslands do produce some orange colors that can be seen in the comparison photos on page 44.

The Present Terrain of Black Hills Agate Locations

The description of locations in the Black Hills including Antelope Spring, S & G, Dow Ridge, Teepee Canyon, and Pringle differ with the exception that they all bear a relationship to the Minnelusa formation.

The Antelope Spring area is a gentle rolling hill area that is covered with trees or open grazing areas. The top soil is shallow with fractured rock lying below the light soil. Some of the rock is exposed and can be seen as the result of forest service roads where soil is graded and exposed. There are some places where erosion has taken place which also exposes rocks.

The Antelope Spring area produced many pounds of material that seemed to be in a small wash that perhaps thousands or millions of years ago had stopped eroding and covered over with soil, then grass, then forests. There likely is more agate in the area if the continuation of the ancient wash can be followed. The examples of Antelope agate can be found on page 44 and page 51.

S & G Canyon is quite different from Antelope Spring. It is similar to the area of Teepee Canyon. There are steep eroded areas where the agate has been exposed by tumbling down from a ridge into the canyon. Some fortification agate has been taken out by actually mining into the limestone. The photo of the S & G Canyon agates found on pages 46 and 52 show such agates.

One nodule that I took from the S & G Canyon is a large piece, 15 to 20 pounds, of solid agate. It is not a fortification pattern piece like Fairburn. If it had been washed into the Badlands, it would qualify as a prairie agate.

One of the Pringle areas is identified by the Forest Service map as "Hand Draw". The eroded pattern of the ridges, when seen on a topographic map takes the shape resembling the fingers of a hand. It is also the area where a mine for Fairburn-type agates was once proposed. It was to be the Springtime Mine. Not a great deal of rock has been moved on the ridge. Fairburns have been found on days I was there in the wash adjacent to the ridge and in situ in the limestone ridge. Quite a bit of exposed agate was found in another wash nearby, but that location still remains unknown to me because Bill has agreed with his fellow rock hounds to keep the exact location confidential for the time being. Examples of the Fairburn material from locations near Pringle are shown on pages 45 and 46. This is not what Zeitner discussed as the material from the Fairhills location. Even though the area is fairly close to Wind Cave National Park, it has not been bulldozed as she described.

The old Fairhills location is not too distant from Hand Draw. The Fairhills agate was taken from near the top of a ridge where the overburden could be bulldozed away. While the main deposit has long ago been depleted, agates are still being found downstream from this location. Burg, who graciously showed me this place, has Fairhill (Fairburn) agates in his collection, several of which are shown on pages 11, 47 and 49.

The agate locations in Custer State Park I have been shown again appear to be along eroding ridges of limestone. Because hunting is not allowed in Custer State Park anymore, I have not actually done any hunting. The State Park agates that I have were supposedly collected before the area was outlawed. In any event, that is what I was told when I purchased these samples. Custer State Park is located east of Custer, South Dakota.

Dow Ridge is another area that falls into the category of remnants of eroding limestone ridges. This area is on private land. Burg has personally hunted the area over many years. He has shown me the area and some wonderful specimens found at that location.

The photo on page 11 demonstrates that Dow Ridge can produce spectacularly beautiful Fairburns.

The mallo agate, mentioned previously in this book, would likely have to come from an area similar to that of Antelope Spring. The Mallo Camp is located just west of the South Dakota border in Wyoming. At this time, nobody known to me can identify the area exactly. My bet is that someone will rediscover it on a search, or by accident, and an additional deposit exposed. The Red Bird Canyon terrain is similar to that of Teepee Canyon.

The Fairburns Are Not All Gone

Finding additional immediate locations may come from looking in tree roots or along the roadways that have been graded by the forest service. One nice agate was found exposed in a forest service roadway while walking on the roads and watching for rocks with pattern. Digging in the immediate area of that agate, however, produced nothing more than a smaller piece of the same agate.

The grasslands and badlands hunting have not been discussed in detail in this book since it is°° not intended to be a trail or guide book. All beautifully patterned and colored agates with the distinctive fortifications that are found in the badlands are automatically considered a Fairburn. One of the primary purposes for this writing was to follow the Fairburn trail into the Black Hills. Although badlands or grassland hunting, whether in places like the original Fairburn beds or Nebraska, is still my favorite.

Zeitner's descriptions of areas to hunt in the badlands are just as valid now as they were 30-40 years ago. The map sent to me by Zieg (see page 17) is still valid as well. The only caution is that many of the landmarks and the faint trails or roads through the grasslands and badlands have changed over the years. If the descriptions of the locations are followed generally, they will lead you to areas which are mostly in the National Grasslands. All these locations yet have agate material waiting to be found. Get yourself maps of the Buffalo Gap National Grasslands, and a Black Hills National Forest, and Oglala National Grasslands map so you can verify public hunting locations. The Black Hills National Forest, the Buffalo Gap National Grasslands and part of the Oglala National Grasslands (northwest corner of Nebraska) provide over two million acres of area to roam and hunt rocks.

Over the next pages, the reader will find agates photographed not only for their beauty, but also as a demonstration of why I have concluded that the Minnelusa formation of the Black Hills produced the Fairburn agate.

The fact that a Fairburn agate may be found in the Black Hills rather than the grasslands or badlands should have no bearing on it's value. Both are equally rare and beautiful and their value is constantly on the increase.

A badlands agate and a Pringle agate. Pringle at the bottom.

Only the upper right agate is from Teepee Canyon. The upper left was found near Pringle. The two at the bottom were found by the author in Cheyenne River gravels near Oral.

One Fairburn and one Pringle. The Pringle (on the right) was not tumbled into the badlands like the Fairburn.

One Fairburn and one Pringle. You tell me which is which.

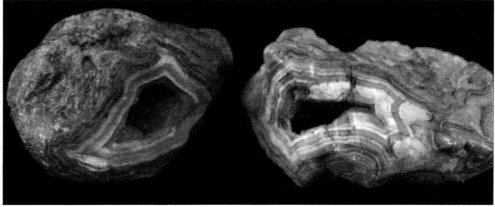

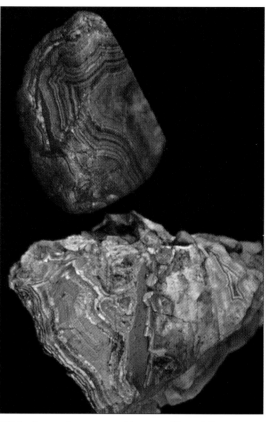

One Fairburn and two Antelope Spring agates. (West side of Black Hills). The ones on the left and top are from Antelope Spring.

This is a close-up of a small Fairburn and an Antelope Spring. The Antelope agate is on the bottom.

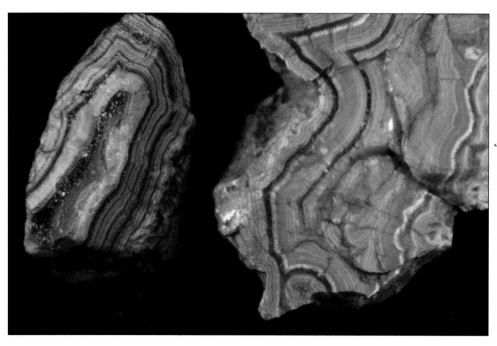

These are examples of the orange color. The agate on the right is from Antelope Spring. The left, a badlands Fairburn.

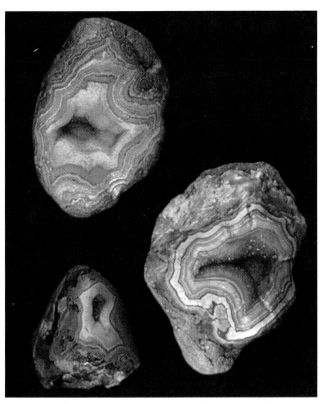

Two badlands agates and one from Pringle. Pringle on the right.

Two Hills agates, one from Antelope Spring and one from the Pringle area. The badlands agate is the one on the right.

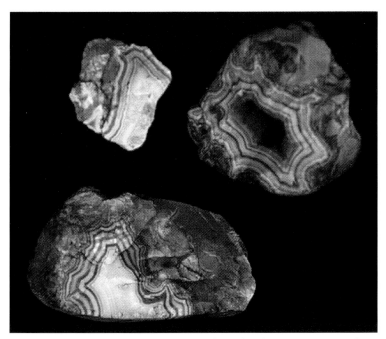

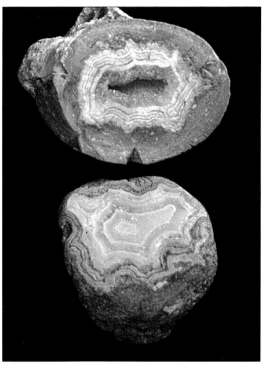

These agates show the great red and white patterns of Fairburn agates. The agate on the upper left, however, was found west of Pringle.

A badlands Fairburn and an agate from S & G Canyon. The agate on the bottom is from the badlands.

The photo shows one Fairburn and two agates from S & G Canyon. The center agate is the badlands agate.

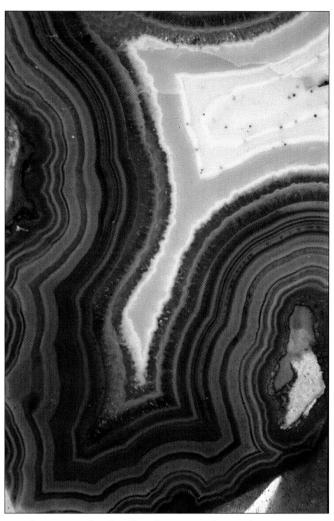

A close-up of a badlands agate from the Bruce Collection.

A close-up of a slab of agate from the author's collection; found near Pringle. A Hot Springs, South Dakota, collector has the rest of the rock from which it was cut.

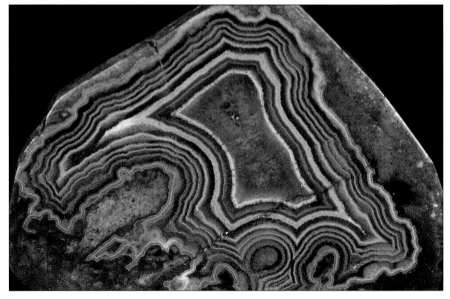

From Burg Collection

When the agates on this page were photographed, it was obvious to me that only the agate hunter who found them could distinguish their location. George Burn identified the location.

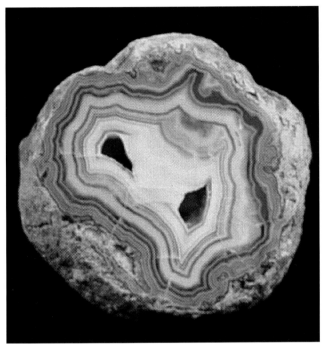 

Agates found southwest of Pringle.

These agates came from the badlands and specifically the Oelrich's beds.

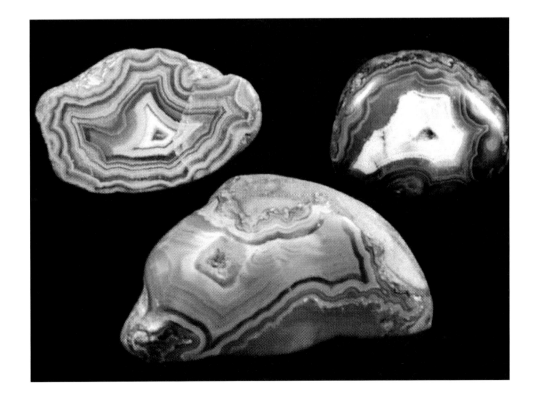

From Burg Collection

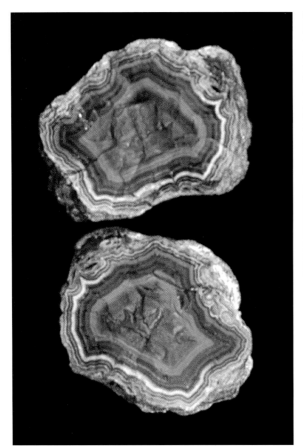

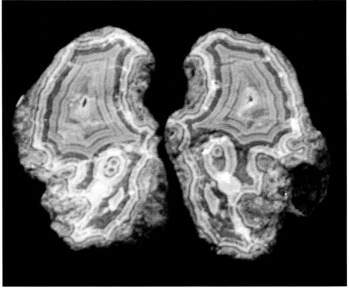

These two pair came from the Minnekata area.

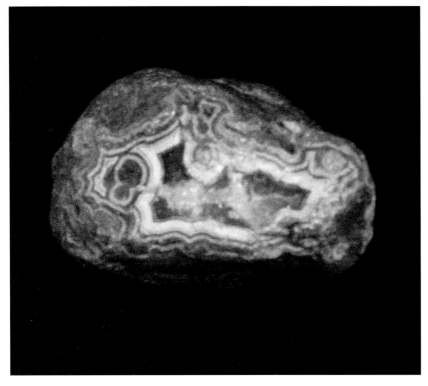

An agate found in Red Canyon north of Edgemont, South Dakota, demonstrates an obvious Fairburn pattern from the western Black Hills.

From Burg Collection

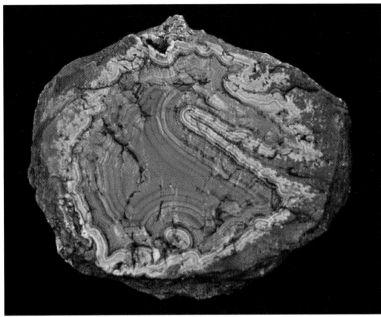

Before seeing these agates, I thought I could identify an agate that came from Custer State Park. Both of these agates were found at Custer State Park. They demonstrate the tremendous variability even within specific locations.

The polished agate below was found near Pringle.

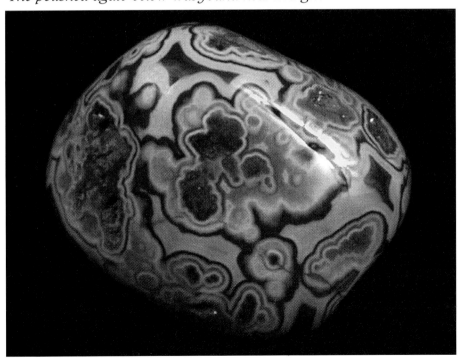

All photos from a slab of Teepee Canyon agate.

It is easy to see the two distinct patterns in this slab. In the text, the theories on agate formations as colloidal systems can account for this occurrence.

These enlargements show how absolutely distinct these two agates are even though they occurred in the same nodule of Teepee Canyon. It's mysterious as to how they could show such different patterns and coloration.

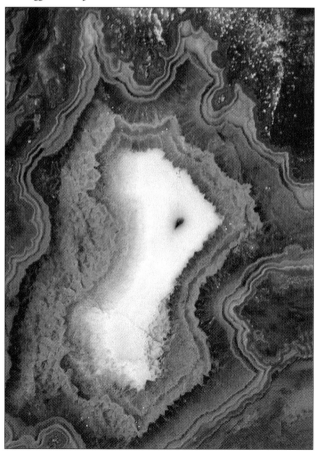

Author's Collection

A striking hills Fairburn from Antelope Spring.

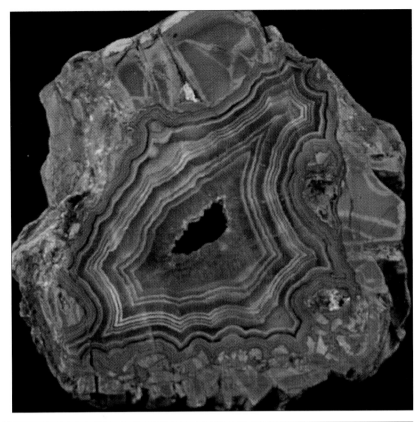

The enlargement of another Antelope Spring Fairburn.

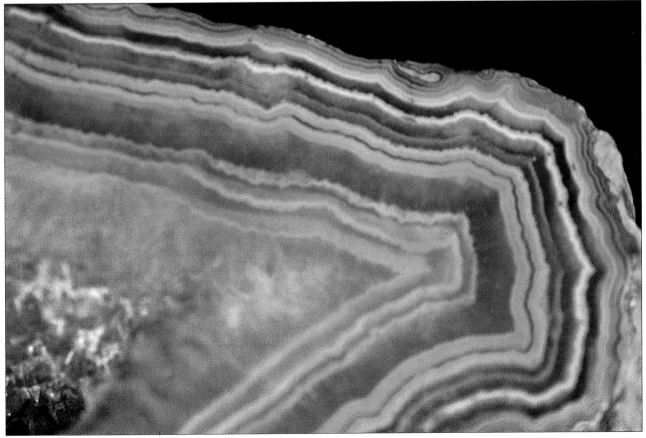

From Seger Collection

The photos to the left and below demonstrate an unusual coloration. They were part of the Louis Miller Fairburn collection acquired by Seger.

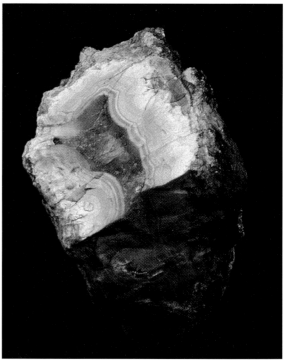

The photo below shows an agate from S & G Canyon that demonstrates the possible origin of the purple agates in the Seger Collection. The agate was uncovered by the author on a 1997 rock hunt.

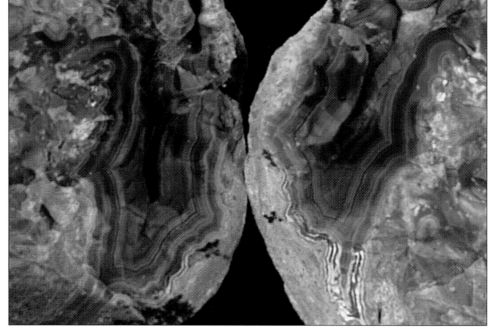

Chapter 6

Black Hills Geology/Badlands Geology

Because the Black Hills and its geologic history have played such an integral part of the Fairburn puzzle, it seems that at least an amateur look at the history is appropriate here.

For more than half a billion years, this area was covered by shallow seas. Estimates vary, but generally agree that between 5,000 and 10,000 feet of sediments accumulated and consolidated into rock. Those sediments, part of which eventually became the formation known as the Minnelusa formation, began to accumulate approximately 310 million years ago. That accumulation accrued over a period of about 50 million years.

The time period began in what is known as the Pennsylvanian and ended in the Permian period between 280 and 230 million years ago. Both the Pennsylvanian and Permian periods occupy the later stages of the Paleozoic era. The following exhibit shows a geologic timetable for further reference. During this period, North America was a part of the ancient continent of Pangea. The land that would become South Dakota was much closer to the equator!

How the Agates Relate to Geologic History

Benjamin Shaub in his book *The Origin of Agates, Thunder Eggs and Other Nodular Structures*, inferred that agates forming in limestone deposits would have formed during the lithification of the host rock. Most studies seem to disagree with that assumption.

The agates themselves probably would not have formed during the Pennsylvanian period when the Minnelusa limestone was laid down. The theories of agate formation do not require the agate formation to occur during the lithification (formation) of the host rock. Agate formation is still the subject of much debate. From current information, however, the most likely process is an accumulation of colloidal silica within the voids in the rock at a later date. The Black Hills formation subsequently experienced conditions favorable to the formations of agates.

The conditions which are currently thought to be necessary for agate formation are discussed in more detail in Chapter 7. External conditions need to be favorable to the accumulation of colloidal silica within the host rock unit. Roger Pabian believes that several factors, including igneous activity, weathering, and volcanic ash, must be present to allow for the development of the colloidal silica. The Black Hills rock units that were eventually eroded into the lowlands were subjected to these conditions.

The agates would then have been formed before the final Black Hills uplift. At this point, there is no other way that I know to estimate the age of these agates.

The geologic events which brought the Fairburns to the badlands is fairly clear even though it

Geologic time scale reproduced from Science *April 14, 1961.*

—with permission of the publisher

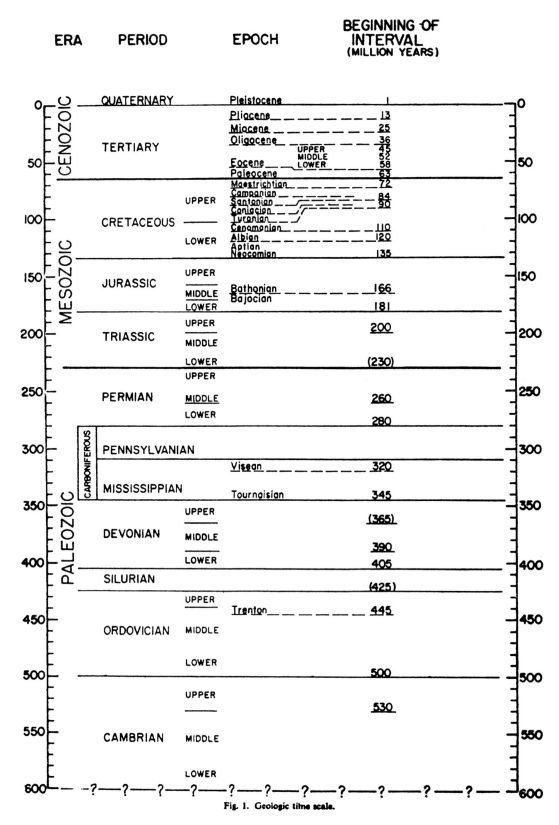

Fig. 1. Geologic time scale.

General Outcrop section of The Black Hills area.

FORMATION			SECTION	THICKNESS IN FEET	DESCRIPTION
QUATERNARY	SANDS AND GRAVELS			0-50	Sand, gravel, and boulders.
PLIOCENE	OGALLALA GROUP			0-100	Light colored sands and silts
MIOCENE	ARIKAREE GROUP			0-500	Light colored clays and silts. White ash bed at base
OLIGOCENE	WHITE RIVER GROUP			0-600	Light colored clays with sandstone channel fillings and local limestone lenses
PALEOCENE	FORT UNION FORMATION	TONGUE RIVER MEMBER		0-425	Light colored clays and sands, with coal-bed farther north.
		CANNONBALL MEMBER		0-225	Green marine shales and yellow sandstones, the latter often as concretions
		LUDLOW MEMBER		0-350	Somber gray clays and sandstones with thin beds of lignite.
?	HELL CREEK FORMATION (Lance Formation)			425	Somber-colored soft brown shale and gray sandstone, with thin lignite lenses in the upper part. Lower half more sandy. Many loglike concretions and thin lenses of iron carbonate.
	FOX HILLS FORMATION			25-200	Grayish-white to yellow sandstone
UPPER	PIERRE SHALE			1200-2000	Principal horizon of limestone lenses giving teepee buttes. Dark-gray shale containing scattered concretions. Widely scattered limestone masses, giving small teepee buttes
	Sharon Springs Mem.				Black fissile shale with concretions
	NIOBRARA FORMATION			100-225	Impure chalk and calcareous shale
	CARLILE FORMATION	Turner Sand Zone		400-750	Light-gray shale with numerous large concretions and sandy layers.
		Wall Creek Sands			Dark-gray shale
	GREENHORN FORMATION			(25-30)	Impure slabby limestone. Weathers buff.
	BELLE FOURCHE SHALE	GRANEROS GROUP		(200-350)	Dark-gray calcareous shale, with thin Orman Lake limestone at base
				300-550	Gray shale with scattered limestone concretions. Clay spur bentonite at base.
LOWER	MOWRY SHALE			150-250	Light-gray siliceous shale. Fish scales and thin layers of bentonite.
	NEWCASTLE SANDSTONE			20-60	Brown to light yellow and white sandstone.
	SKULL CREEK SHALE			170-270	Dark gray to black shale
	FALL RIVER [DAKOTA (?)] ss	INYAN KARA GROUP / LAKOTA FM		10-200	Massive to slabby sandstone
	Fuson Shale			10-188	Coarse gray to buff cross-bedded conglomeratic ss, interbedded with buff, red, and gray clay, especially toward top. Local fine-grained limestone.
	Minnewaste ls			0-25	
				25-485	
JURASSIC	MORRISON FORMATION			0-220	Green to maroon shale. Thin sandstone.
	UNKPAPA SS			0-225	Massive fine-grained sandstone
	SUNDANCE FM	Redwater Mem / Lak Member / Hulett Member / Stockade Beaver / Canyon Spr. Mem		250-450	Greenish-gray shale, thin limestone lenses. Glauconitic sandstone; red ss. near middle
	GYPSUM SPRING			0-45	Red siltstone, gypsum, and limestone
TRIASSIC ?	SPEARFISH FORMATION			250-700	Red sandy shale, soft red sandstone and siltstone with gypsum and thin limestone layers. Gypsum locally near the base.
	Goose Egg Equivalent				
PERMIAN	MINNEKAHTA LIMESTONE			30-50	Massive gray, laminated limestone
	OPECHE FORMATION			50-135	Red shale and sandstone
PENNSYLVANIAN	MINNELUSA FORMATION			350-850	Yellow to red cross-bedded sandstone, limestone, and anhydrite locally at top. Interbedded sandstone, limestone, dolomite, shale, and anhydrite. Red shale with interbedded limestone and sandstone at base
MISSISSIPPIAN	PAHASAPA (MADISON) LIMESTONE			300-630	Massive light-colored limestone. Dolomite in part. Cavernous in upper part.
DEVONIAN	ENGLEWOOD LIMESTONE			30-60	Pink to buff limestone. Shale locally at base
ORDOVICIAN	WHITEWOOD (RED RIVER) FORMATION			0-60	Buff dolomite and limestone
	WINNIPEG FORMATION			0-100	Green shale with siltstone
CAMBRIAN	DEADWOOD FORMATION			10-400	Massive buff sandstone. Greenish glauconitic shale, flaggy dolomite and flatpebble limestone conglomerate. Sandstone, with conglomerate locally at the base.
PRE-CAMBRIAN	METAMORPHIC and IGNEOUS ROCKS				Schist, slate, quartzite, and arkosic grit. Intruded by diorite, metamorphosed to amphibolite, and by granite and pegmatite.

1963

left a confusion of stratigraphic units. About 65 to 70 million years ago a slow uplift began bringing the dry land to the interior of the continent. In the *Roadside Geology of South Dakota,* Gries paints a picture of what happened after the uplift began.

> *When the uplift started, the land surface was close to sea level, and the rocks which form the top of Harney Peak lay 7,500 feet lower, so that minimum uplift was probably on the order of 15,000 feet. Erosion of the newly uplifted area started immediately. It was rapid at first when the soft Cretaceous shales were exposed, but slowed down materially when the older, more resistant limestones, sandstones, and volcanic rocks were being removed.*

> *As the uplift proceeded, huge volumes of molten rock forced their way into the existing rocks in the northern Hills, expanding their volume and forming smaller domes upon the larger one. Deposits of gold, silver, lead, zinc, and tungsten were formed at this time.*

> *Erosion continued after the uplifting ceased, and by roughly 37 million years ago, the Black Hills had much their present size and shape. Hard rocks formed the ridges and plateaus; softer rocks eroded into valleys and basins. From the air, the Hills would have resembled a huge, oval target, with the oldest rock forming the bull's eye, surrounded by successive rings of younger rocks.*

> *Erosion of newly uplifted Rocky Mountains to the west initiated yet another cycle of deposition in this area. Debris from the higher Hills, and light-colored clays and volcanic ash from farther west was no longer swept away by streams and winds, but began to accumulate on the old erosion surface. Much of this was deposited in what is now the Badlands, but the sediments lapped up on the Black Hills until probably less than 2,000 feet of ancient rocks stood as a ridge above a broad, featureless floodplain.*

> *Fickle nature again reversed herself, and started unearthing the Hills by eroding the light-colored clays. That task is now fairly well completed in the higher Hills, but remnants of the clays and gravels remain up to an altitude of at least 5,300 feet; thick blankets of light clays and gravels remain on the stream divides east of the Black Hills, and in the Badlands area.   —page 214*

It has previously been noted that the exposed portion of the Minnelusa formation is found as a ring of limestone around the Black Hills. The Geologic Map of the Black Hills found on pages 36 and 37 clearly demonstrates that limestone plateau.

The Minnelusa Formation as Seen in the Black Hills

The Minnelusa formation has various descriptions that characterize it throughout its thickness which ranges from 75 to 1300 feet in thickness.

Yellow to red cross-bedded sandstone, limestone and anhydride locally at top.

Interbedded, sandstone, limestone, dolomite, shale, and anhydride.

Red shale with interbedded limestone and sandstone at base.

Because it is so massive, in places where it has not been eroded away, it can be seen as rock walls lining such canyons as Redbird and Hells (Teepee) Canyon.

The uplift of the Black Hills and the location of the Minnelusa remnants play an important part in the eventual location of Fairburns and their distribution east of the Black Hills.

Erosional Forces Directed Eastward

The geology of the Black Hills as it relates to stream flow is an important piece of the puzzle. Fairburns can be found in Teepee Canyon and Redbird Canyon, S & G Canyon, and Antelope Spring. All of these locations are toward the western side of the Black Hills uplift. Why then did the Fairburns not distribute to the west as well as to the east?

The map taken from the *Geology of the Black Hills* shows the general geology of the Black Hills. It also shows the many streams and rivers from the Cheyenne on the south to the Belle Fourche River on the north that eroded the Black Hills almost totally to the east toward the Badlands which themselves were formed in part by stream debris. (See page 27)

Previously in Chapter 5, it was demonstrated that Teepee Canyon agate shows up in the gravels of the badlands to the east. Another look at the geologic map of the Black Hills (pages 36 and 37) demonstrates the large Cenozoic (less than 65 million years) deposits that accumulated easterly from the Black Hills.

There should be no mystery therefore, about why these varied and scattered accumulations of Fairburns ended up in the badlands and grasslands east of the Black Hills. Those sediments which are now being eroded and exposed in the badlands are the same sediments that would have originally eroded from the Black Hills between 70 and 35 million years ago.

The confusion of their distribution would have been enhanced when they began to erode again in recent (less than 30 million years) geologic times. The badlands formations that are now so stark and rugged are very recent in geologic time. They are now eroding away two previous depositions and further complicating the badlands geologic puzzle.

There are millions of yet unexposed Fairburn agates waiting to be brought to the surface. If we were, somehow, through a time machine, able to cause an immediate erosion equal to one hundred years, rock hounds would again be delighted to find numerous Fairburns exposed on the surface.

There are also vast areas of the Black Hills that undoubtedly will produce Fairburn agate material. A large part of the area is Black Hills National Forest and is open for hunting.

While not all will agree that agates from the Hills can be called Fairburns, my conclusion is that they are the same material from the same geologic formation. My preference is to refer to Fairburns as being either badlands Fairburns or hills Fairburns, depending upon where they are found.

Chapter 7

Agate Genesis and Genesis of Fairburns

Although satisfied that the Black Hills limestones were the birthplace of the Fairburns, there was still a curiosity about why and how these fortification agates were formed. Why were some of them distinctive as to location? Why were there variations within individual locations? Were they all formed the same way?

When I began agate hunting, I accepted the standard explanation of agate formation. This standard explanation normally consisted of some form of infiltration or leakage of silica materials into a cavity or pocket in the rock.

When I began contemplating the origin of the Fairburn agates and putting that theory in some form of writing, it became clear that it was difficult to use standard explanations. Having mined Teepee Canyon and Dryhead agates, it was obvious to me that these beautiful agates were formed in cavities in limestone rock, but equally obvious was that there did not appear to be any form of supply of agate material (i.e., channels) that could explain how the material could be deposited deep in the limestone. If one were to assume that this silica agate was forming in the cavity, why would it not be forming in channels or areas of infiltration around the rock?

The leakage part of the theory did not seem to make sense because there would have to be other evidence:

1. How could the agate be isolated in limestone without evidence of seams or $SiO2$ channels?

2. Why was there no stratigraphy of the agates or materials nearby? By this, I mean that each agate should have the same band or color sequence if, in fact, the material flowed into the area and simultaneously filled these cavities from the outside.

I read what was readily available for agate theories. Nothing gave me, with any convincing evidence, a reasonable explanation. Liesegang had been used as the standard to show that the coloration of agates arose from the pigmentation which somehow assisted in the creation of the rings or banding in the agate. These Liesegang rings then had been the explanation for banded agates. Liesegang had demonstrated using a silica gel rather than a leakage or duct theory. No explanation was given for the "leakage" channels in these fortification agates.

More Questions Than Answers

About ten years before my renewed interest in rocks, John Sinkankas wrote an article that was published by the Lapidary Journal in the book *The Agates of North America*. It was one of my early acquisitions of written material.

The Sinkankas article was entitled *What Do We Really Know about the Formation of Agate and Chalcedony?* The article raised the same questions bothering me.

1. *In every kind of rock, the question of "where did silica come from" is a good one to ask, particularly in the case of basalt, which is noted for being very poor in quartz as a constituent mineral, yet is often rich in chalcedony in seams, veins, and nodules.* —page 7

2. *As for the "canals" many nodules freshly dug out of basalt show absolutely no trace of chalcedony pipes or canal continuations which lead away from the nodules into the surrounding rock.* —page 8

3. *Why doesn't the cavity fill up with the silica solution and then stop right there? Why should it keep attracting more and more silica until it is sometimes filled completely? I cannot answer these questions except to say that the evidence of the nodules themselves shows that they do absorb successive waves of silica solutions...* —page 8

Sinkankas went on to conclude that "the problems of explaining the formation of chalcedony, its inclusions, and its colors, need more study before thoroughly convincing explanations can be advanced. It is regrettable that more qualified mineralogists have not taken up the challenge offered...." As you will become aware as this book progresses...others have now taken the challenge.

How then does one explain the origin of the agates? The standard theory was that SiO2 (silica) had filled the cavity with successive layers. As a result of the layering of successive solutions, it was speculated the ducts or leakages could be seen in the agates. An example of these types of alleged leakages are shown in the Fairburn agate on page 22. This theory then was supposed to account for an agate found in a basalt or limestone cavity.

It must be pointed out that when Fairburn agates are found in the badlands, many of the larger ones show pattern in a matrix of jasper or chert. The same jasper or chert matrix is found around the Fairburn fortifications in the hills (Teepee Canyon and others) agates. The quartz or silica has clearly permeated the limestone immediately around the agate itself and has formed a jasper or chert nodule in which the patterned agate is encased. The same thing occurred with "Currington" agate you can identify the brown matrix. (See photo on page 61).

Zeitner speculated briefly in her accounting about the duct theory that left leakage channels in and out of the agate. This, she proposed, would be an excellent explanation why you could get such a wonderful variety of color in the Fairburn. There was no discussion of the origin of the jasper or chert matrix.

Questions at Teepee Canyon

The old theories had sounded somewhat plausible, until I mined for Teepee Canyon agate with Ray Currington. He was an old agate hunter who, at that point, had a mine claim which produced Teepee Canyon agate. He was selling what he produced to local rock shops and other collectors such as myself. He had some Fairburn agates including one which after being tumbled, weighed about 12 pounds. (See photo on page 61). I have heard it rumored recently that the agate sold for about $200 a pound. It was a beautiful agate and certainly worth it, if you had the money to spend.

The Teepee Canyon agates at Currington's mine were coming out of what I later learned was the

A "traditional" Teepee Canyon and a Fairburn demonstrating matrix.

The group of agates shows the variety of color and pattern that can come from Teepee Canyon.

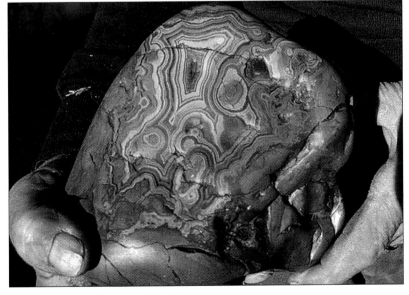

Ray Currington's 12-pound Fairburn.

Minnelusa formation. The agate is mined by removing the overburden and then physically separating massive layers of limestone that can be from a foot to 16-inches thick. Currington did all this with chisels, pry bars and jacks.

The nodules could be seen in the host rock at areas where the layers separated. Sledgehammers break them loose. They are then broken or cut open to identify whether or not they do contain the fortification agate called Teepee Canyon.

Notice the soft surface on these layers and notice that the outside surface of the nodule is not compact. Rather than being a hard surface it is of a powdery nature which shows on the hands after rubbing on the surface. (See photo on page 63)

Inside, however, the agate, and the matrix around it are well-formed and solidified into stone the hardness of agate, jasper, or chert. How could this cavity be full of agate when the limestone rocks around it maintained the limestone character even to the point of being a powdery limestone rock? The leakage theories proposed that the silica had leaked through these layers and leaked into this cavity. There was, however, not a trace of any leakage, a duct, or any trace of silica or solidified material. Neither was there any evidence of a leakage on the outside surface of the nodule. It was obvious, in this instance, that there was no trace of a duct or channel that had leaked in or out of the formation around the nodule.

The problem of trying to explain the beautifully fortified and colorful agate in the middle of a limestone layer brought back the same question again. How do agates form? The explanations were not satisfactory to explain my field observations.

Earlier Contact with "Limestone" Agates at Dryhead Canyon

Years before mining at Teepee Canyon, I had been to the Dryhead agate mine in Dryhead Canyon, Montana. Dryhead agate is also found in a type of a limestone formation. Those nodules were also formed isolated in pockets or cavities in soft limestone layers which, other than the agate, are not permeated with silica. While you can find evidence on the surface of some of the nodules that

Above: Ray Currington lifts a slab of limestone with a jack. Right: An overview of the Teepee Canyon agate mine site gives some indication of the tons of material moved using only hand tools.

The nodules showing on the top side of this slab in Minnelusa limestone would be broken loose with a sledge hammer. They would then be cracked open or sawed to see if they contained Teepee Canyon agate.

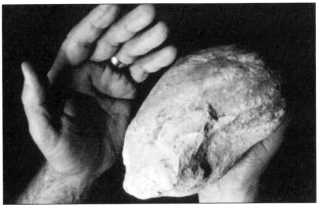

The surface of the nodules was so soft and powdery that it came off on the hands. There was no indication of the beautiful fortification agate that filled the center of the nodule.

contain agate, there are many nodules showing no evidence of ducts or leakages within or around the agate. At that point in time my knowledge of agate formation was so limited that I had not yet determined to pursue my questions further.

In the early 1990s, I ran across the book, The *Origin of Agates, Thunder Eggs and Other Nodular Structures* published in 1989 and written by Benjamin Shaub. Shaub proposed some new theories which stimulated my interest in pursuing an acceptable theory of agate formation.

His theory was that agates resulted from agglutinated colloidal silica. This new theory would explain some questions, such as the fact that the infiltration channels or leakage openings were really not a realistic explanation of how the agate formed. The theory explained that the leakage channels were formed by a deformation process pushing material **out** of the cavity rather than draining material into the cavity.

These explanations seemed more logical than others I had heard for the leakages. The book, however, was not written in terms of a chemical or mineralogical analysis which would allow for a chemical understanding of how the silica could reach the pocket or how the agate banding or other structure within the pocket took shape.

Because of Shaub's theory, my research turned to investigating colloidal silica. Needless to say, colloidal silica is seldom addressed in laymen's terms in the chemistry book. The *Dictionary of Geologic Terms* defines silica as:

> *Silicon dioxide, SiO2. It occurs as crystalline quartz, cryptocrystalline chalcedony, and amorphous opal, dominantly in sand, diatomite, and chert; and combined in silicates as an essential constituent of many minerals.*

It was clear that an understanding of how silica behaved chemically was going to be essential. Since I have no chemistry background, the literature which I was able to find was basically incomprehensible. I needed answers from someone who knew geological chemistry, especially as it related to silica.

A number of inquiries were made at universities and at the School of Mines, but there was very little interest in this very common form of mineral, or agate.

How was I going to find out whether or not the silica could behave in the way Shaub described it? I needed answers to all those questions that had not been explained. I needed to satisfactorily address problems and difficulties with the theory that continued to bother me such as:

1. Why some agates were banded and some were not?

2. What actually caused the banding? (Liesegang said it was the pigmentation.)

3. How did the silica travel through the rock and get into the cavities of the sedimentary rocks such as those containing the Teepee Canyon agates?

4. What pressures or temperatures would allow for agate formation?

5. Why, even with agates such as Teepee Canyon, dug out side by side, was there no stratigraphy? (Identical or corresponding banding.) (See photos on page 50.)

6. Even though the color characteristics are similar in agates such as Teepee Canyon and Dryhead, they clearly vary in design, color, and banding. Why was that?

7. How did the "fortifications" in agates such as Teepee Canyon and Fairburns come to be created?

8. What controlled the fortification or bandings within the agate? (External forces, i.e. leakage, or internal chemistry).

9. How could agates form in pockets in sedimentary rock without having any channel or an effect on the host rock other than immediately adjacent to the agate?

These questions had begun to arise when early hunting excursions took me to the shores of Lake Superior. When examining the basalts that have been identified as the host rock of the Lake Superior agate, you can find seams of agate-type material. These seams, however, normally do not connect the pockets of agates that have developed in the basalt. In addition, when you find a pocket filled with agate material, the pockets around it may not be filled at all. Even if the pockets are filled, they are not consistent either in texture, color, or pattern, with the agate from the pocket first examined.

Chemical and Mineralogical Search Begins to Prove Fruitful

My search for an understanding of agate genesis lead me to the library. A computer search produced only one recent article that appeared to address the issue of these problems. It was written by Michael Landmesser, was in excess of 125 pages in length, and it was written in German! The abstract of the article, however, was extremely intriguing. It indicated that the author had analyzed all of the old agate theories, and further demonstrated their errors and proposed the fundamentals of a new agate theory. I could not, however, read it. Translations of technical papers are very expensive. It sat in my file for years before I returned to it.

I returned again to Shaub and his book *The Origins of Thunder Eggs and Other Agate Materials*. At least it contained the first logical explanation of agate formation that I have discovered.

The theory proposed agates formed when agglutinated colloidal silica came together in the rock formation. He had explanations for the variations in the agate structures (i.e. moss versus banded), but the book was still lacking in chemical and mineralogical analysis to support the theories that were proposed.

I decided to talk agates with Shaub. I tracked down his home through the publisher. Unfortunately, when I called, his widow answered. I was six months too late.

The article of Landmesser was still in my possession. The abstract of the article talked of a new theory which sounded very similar to that of Shaub. The abstract also said the Landmesser had examined the agate theories and was proposing a new theory involving colloidal silica.

It was obvious to me that the only answers I was to get would come from an interpretation of that article. A search began for libraries throughout the United States that would not only have the article, but might possibly have a translated version of the article.

I could find only five libraries in the United States that even had a copy of the article. It was not available in English. My only alternative was to turn to a translation, which for my rock hound hobby, was a very expensive venture.

My determination to find out about the new theory in the German article resulted in my contacting a local university (Lawrence University in Appleton, Wisconsin) for translators. Lawrence University has an excellent German department and I did find a very capable translator who did the eventual translation. It was a difficult translation because it is a technical scientific work. As it turns out, however, the interpretation and translation was well worth the investment.

Landmessers *The Problem of Agate Genesis* is a very technical publication, adding a wealth of knowledge to the field. As far as I know at this time, it has not been published in English. Should any reader be aware of a university or other technical publisher who may be encouraged to publish the work, please let me know. The translated version would be made available for such a venture.

Chemical and Mineralogical Analysis Can Provide Answers

There was a comprehensive analysis of old theories. The explanations as to why those theories did not work was very helpful. The new theory explains from the chemical standpoint why the others failed. Dr. Landmesser is a highly qualified mineralogist with the technical expertise to analyze this complex process.

As soon as the translation was available to me, I knew it was important to see the original photography that came with the original article. Once I had purchased an original from Germany, I could examine the text as well as the illustrations for explanations of this new theory. The theory is premised upon the accumulation of colloidal silica within the host rock. An explanation of the theory in greater detail is contained in Chapter 7. The theory also explains the variability of agates that result from the erratic behavior of colloidal silica when it makes the transition to what we call agates.

The new theory explains why and how, from a chemical and petrological standpoint, agates form. It explains why some agates are banded and some are moss. It explains how some form in volcanic rock and others in limestone. As you will see in Chapter 7, it helps answer questions that allow for an explanation of the genesis of Fairburns.

The Fairburn Connection

Now I had information that would allow for the similar origin, and tremendous variability of Fairburns. There was now no question in my mind that Fairburns, State Park agate, Teepee Canyon agate, Antelope agate, S & G Canyon agate, and Pringle agates, likely could and did form in the Minnelusa formation. The connection can now be made because the multiple places where they can still be found in the Black Hills have a common connection. These locations are all in or associated with the Minnelusa formation.

The fortification structures and delicate banding with multiple colors in conjunction with clear quartz crystal layers group them into a similar classification of agates which are structurally and colorfully similar. Yet local variations in trace minerals caused a multiplicity of colors. The variability of the individual colloidal systems translated to a multitude of patterns.

Zeitner had mused that the Fairhills agates might be related to the Fairburns, but was not willing to make the direct connection. Art Bruce refused to make the connection to the limestone agates (his reference to Teepee Canyons). Our knowledge, however, of the geology of the Black Hills and badlands dictates that alluvial deposits now found in the badlands came from the Black Hills. In like fashion, the Fairburn-type agates that are found in the badlands may also be found yet in the Black Hills. The formation, known as the Minnelusa formation originated over 260 million years ago and eventually became the host in which the Fairburn agates of southwestern South Dakota were formed.

Rock hunters who explore the grasslands, badlands, riverbeds, or gravel quarries located in the riverbeds, will find many varying materials including fossil bones, cycads, petrified wood, chalcedony, and other forms of agate. It is the jumbled geologic history that has caused these materials from many different origins to be found together.

There are materials which actually originated in the badlands after the badland formations were created. Some of the fossil bones and turtles originated there. The float chalcedony originated in the badlands, but the agates of the Fairburn-type formed in the Minnelusa formation which once covered the Black Hills.

The cycads which are found in the Black Hills and badlands came on the scene at the Triassic time, beginning approximately 230 million years ago. Because they were on the scene for millions of years, it is unclear whether the cycads were petrified at the same time as the Fairburn agates were formed. They have, however, along with petrified woods and agates been tumbled together and can be found scattered throughout the grasslands and badlands areas. Hunters are just as likely to find a Fairburn at a gravel pit as a petrified cycad. I am personally aware of a large (over 20 pounds) piece of cycad recently found at a gravel pit along the Cheyenne River.

There is no way, of which I am aware, to isolate a time period when the petrified ferns and cycads and other petrified woods could have been formed and later deposited in the grasslands and badlands. Because of the way the Black Hills geologic "blister" formed, it caused over a period of more than 30 million years, tremendous erosional forces. The uplift tumbled together over a mile depth of various rocks eroded off the top of the Black Hills as they rose. There is estimated to be 7,000 feet of material missing which was originally over the top of the Black Hills. Those materials are the ones which are found in the alluvial deposits of the badlands and grasslands.

Chapter 8

The Riddles of Agate Formation/Agate Formation Theories

Throughout this chapter reliance for my interpretations is primarily upon the works of Landmesser and Pabian. Their work will be discussed in greater detail later in the chapter. Their works are referenced in the Bibliography.

There now seems to be general agreement that the SiO2 solutions accumulate in colloidal form at a given point. Depending upon the local conditions, the agates take on various internal structuring and coloration.

Some agates are banded and others take various forms like moss. The striking feature of the Fairburn agates, of course, is the banding. The banding is likely the result of internal rhythms. It is believed by Landmesser that the decisive structure determining or outlining the process during the ordinary banding is the formation of what is called a spherulite. Those who have closely looked at the Fairburn agates from the Black Hills and badlands have undoubtedly noticed that many times you can see the fabulous Fairburn fortifications arise from spherulites that line the cavity. (See photo on page 14.) The pattern of these spherulites varies, but in the best fortification agates they are very clear.

The Fairburn is especially known for its delicate fortification patterns. The conditions must have been excellent for the formation of spherulites which dictate the pattern result.

The optically visible lines and patterns that are so fascinating **do not result from the pigmentation** as Liesegang surmised. He had proposed that it was the pigments that caused the bands to form. Landmesser, however, shows, in fact, that the Liesegang rings can form at opposite angles to the true agate banding. Landmesser believes that he can substantiate that the fine banding of the chalcedony depends upon the periodic fluctuation of the density in the agate. These fluctuations he proposes are caused by a periodic change in the size of the subparticles and the pore spaces.

Another question has always been how agates can have so much variety. Some colors are similar, but Fairburns are like snowflakes, every one different. Yet they developed basically from the similar solution in close proximity to each other. In fact, you can see on page 50 that at least two distinct Fairburn patterns and colors have developed in what became one solid rock. Although I call them Fairburns, be aware that this rock actually came from the Black Hills in the Teepee Canyon area.

Landmesser's agate theory also addresses the internal variability in *The Problem of Agate Genesis:*

> *...it happens that one achieves different results in the case of colloids under 'apparently' the same experiment conditions whereas all the properties of even-tempered salt solutions are, for example, the same...Every solution and every colloid system appears as a single individual; all of its properties and its changes are individual, its entire life is individual.*
> *—page 111*

These figures are the author's rough sketch of a proposed sequence in the formation of an agate such as a Fairburn.

(1) Silica permeates the rock and accumulates as colloidal silica.
(2) Spherulites form first on the rock surface.
(3) The pattern forms as the silica precipitates over the spherulites.

1.

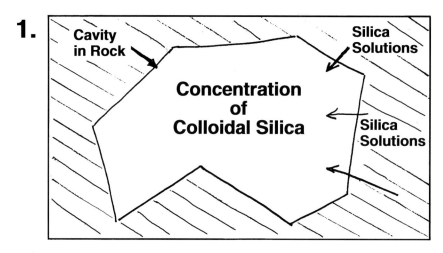

2.

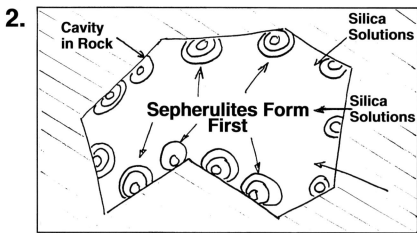

3.

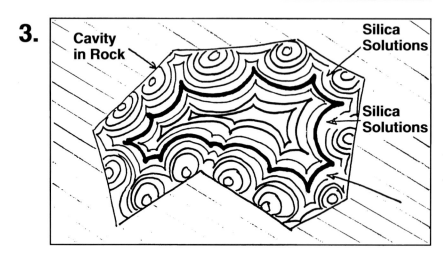

Add the chemical and trace mineral variability to the colloidal system and the stage is set for the fantastic variety. That is so even in Lagunas and Lake Superiors. Lake Superiors are known to be more limited in color, but yet no two Lake Superiors are the same. Each colloidal system varies like a snowflake.

When Did the Fairburns Form?

As indicated previously, it is likely that the Fairburns were not formed during the lithification processes of the Minnelusa. When the actual formation took place is still a mystery. Was this area a silica-rich warm spot millions of years before the actual volcanic processes uplifted the Black Hills? The theories of Pabian would propose that it is the ash layers that provide the silica-rich waters allowing for the agate formation. The Black Hills area was covered with volcanic ash from eruptions to the west when the Rocky Mountain chain was forming.

New agate theories propose an analysis that not only makes sense from the technical point of view, but also makes sense when put in practical application by the rock hound. The new theories propose the formation of various agates as well as petrified materials as I have summarized them in the pages that follow.

Roger K. Pabian of the University of Nebraska has proposed theories that in many ways appear similar to those of Landmesser. Those summaries and excerpts of those proposals are also set forth later in this publication. It appears that primary distinction between the two scholars is the issue of internal forces versus external forces as they dictate the development of the structures.

A. Agate Genesis: Landmesser

Michael Landmesser's publication, *The Problem of Agate Genesis,* is a very comprehensive review of all of the previous agate theories that had been proposed. There is an excellent technical and chemical analysis of each one of the theories. It is a very scholarly work. It has not yet been published in English, but as indicated, it has been translated.

In his publication (published in 1984), Landmesser summarized the problem that has faced us as hobbyists when trying to explain about agates to others:

> *The problem of agate genesis has been discussed very controversially for more than 200 years up to very recent times. One isn't even in agreement in the rough about the formative conditions of agates; theories of agate formation in the magmatic, post-magmatic and the sedimentary realm oppose each other. At the same time, an overall presentation of the agate question is missing in the literature in which the different theories, objections, and relevant observations in nature are concisely formulated.*

Landmesser, in his book *The Problem of Agate Genesis,* has done a comprehensive work dealing with the historical problem of banded chalcedony aggregates.

Landmesser begins his analysis with the early work of Wilhelm Haidinger (1848) on the metamorphosis of rocks. The early theories of Haidinger, Noeggerath, and Kenngott generally involve volcanic activity and infiltration or deposition into cavities from external sources. There was even a geyser or spring theory. Part of another theory by a writer by the name of Brachmeyer proposed that the agate-forming solution stemmed from humus soils of tropical forests.

All of the older agate theories usually assume the existence of open fissures to allow solution movement and deposition.

As Landmesser points out, however, those who have found agates in their place of origin, realize that there is no, or seemingly no, stratigraphic relationship between the cavities. Lake Superior agate in the basalt can be immediately adjacent to another type of formation. The same variability in agate nodules has been observed by myself in Dryhead Canyon and Teepee Canyon.

Landmesser deals further with the Liesegang theory with which most rock hounds at some point in time have become acquainted. The theory proposed the banding in fortification agates resulted from the formation of pigment rings in silica gel. These pigment rings were the Liesegang rings. It was the formation of the rings which concluded the first stage of the agate genesis. Landmesser concludes that the Liesegang rings do not bear out in close scrutiny in the actual setting, as Liesegang had proposed.

Theories of volcanic and post-volcanic agate formation along with numerous other theories are examined in this comprehensive work. The conclusions of Landmesser, however, result in his proposal for a new agate theory. Following the comprehensive analysis of reasons for his new theory, Landmesser proposes in *The Problem of Agate Genesis* as follows:

> *In conclusion, we arrive at the following result: agates are cavity in fillings in rocks which originated in the sedimentary-digenetic P-T (pressure-temperature) realm, i.e., above 0° C and below 200° C. Agates originated from poly-dispersal colloidal systems, a process in which the pressure curve is not surpassed. Internal rhythms play a decisive role in their formation. Agate zones with ordinary banding have, as a rule, passed through a gel stage. The decisive, structurally determining process in the development of ordinary banding is the formation of spherulites. The structure of the "infiltration channels" arises through a deformation process in the gelatinous media.* —page 126

He proposes that the central processes of substance transport during the genesis of agates is a "diffusion" process. Open supply paths to the cavities are not necessary. The silicic acid is transported mainly in what is known as a "monomere" form. In the course of the genesis of agates, especially however during the crystallization of SiO_2 in the originating agate, the agate is continuously supplied with SiO_2. Landmesser proposes that the origin of agates and the salicification of woods are related processes.

In his prelude to proposing his theory, Landmesser deals with several aspects of agate formation, thereby explaining why he comes to his conclusions:

1. There has to be a supply of SiO_2 (silica) during agate genesis and the best candidate for that is the diffusion of monomere silicic acid.

2. Colloidal systems, in their formation, behave differently even under strict laboratory controls. Landmesser concludes that **every colloidal system appears as a singular individual**; all of its properties and its changes are individual, its entire life is individual. This explanation, for me, helps me to understand why no two Fairburns are completely alike and why it has been so difficult to track down an origin of the Fairburns.

3. It is the internal rhythms of the colloidal systems that explains the multiplicity of structures.

4. The transportation of silica to the banded agate or chalcedony has been observed naturally and in the laboratory within the temperature realms proposed in the theory.

What about the genesis of the Fairburn then? How is it explained by the new theory described in short as polydispursed colloidal systems resulting from silicic acid transported to the forming or developing agate in monomere form. That is a mouthful for saying that the new theory states that agates are former colloidal (silica) systems, which originated from a supply of silica (in monomere form) which permeated the host rock.

Landmesser explains that the silica had to be transported through the rock to the place where the agate nodule would form. It is the monomere form of the chemical which allows for the silicic acid to be transported without overall affect on the adjacent rock. It is important to note that this theory does not cause the silicic acid to pass through channels, but rather to be dispersed through the rock. The fact that this dispersion can happen without making chemical changes to the rock formation is extremely important to the understanding of how these nodules could come together.

The silicic acid then becomes a colloidal accumulation precipitated from the solution. Landmesser deals with the fact that these agates must form in a limited pressure and temperature realm.

It is also important to note that this process occurs between 0° centigrade and 200° centigrade. Agate material in this form would likely not form above or below that range. Such conditions would likely occur after the rock itself had formed because the agates themselves form in pockets or spaces within the rocks.

The solutions are continually supplied because as the solution moves from a liquid to a colloidal gel and then to a hard silica, there is a loss of volume of fluid from which the silica has precipitated.

Landmesser talked about the temperature range for the formation state, but had not yet established a pressure range. Although he had not yet completed analysis of the pressure range, he speculated that it was likely fairly low.

Landmesser has also published analysis about the petrification of wood. *Lapidary Journal* did a feature article in the August 1995 issue, summarizing the theory which deals with the manner in which wood is petrified by silica (SiO_2).

As the years go by the rock hound world will undoubtedly hear more from Landmesser, who is referred to as a "brilliant young geologist/mineralogist". Landmesser is located at the Institute for Geological Science, at Gutenberg University in Mainz, Germany.

B. Agate Genesis: Pabian

Roger Pabian, of the University of Nebraska, has published several papers on agate genesis. The most recent is the 1994 publication *Banded Agates, Origins and Inclusions*. He has attempted to correlate how the banded agates originate when considering the "technical, stratigraphic, and geochemical conditions that must be present before an agate can form."   —page 3

Pabian appears to deal with a silica gel accumulation similarly to most current writers. However, he believes that the chemistry external to the silica gels plays a significant role.

There also appears to be a significant difference between the conditions leading to the formation of marine sedimentary agates, (such as Fairburns) and those agates formed in basalt, (such as Lake Superiors). Eventually, however, the preconditions must generate a source of silica.

In his most recent publications, Pabian is much more specific about conditions leading to the eventual agate formation. As a part of the discussion, Pabian outlines the fact that there are numerous formations within which banded agates occur and he listed five basic lithologies (rock types) in which they are found: (1) Thunder eggs that occur in rhyolitic, welded-ash flow tuffs. (2) Amygdaloidal agates that occur in vesicular theoleitic basalts and their silica source derived from overlying rhyolitic ashes. (3) Amygdaloidal agates that occur in vesicular andesitic rocks. (4) Nodular or veined agates that form in continental sedimentary rocks such as claystones. (5) Nodular agates that form in regressive marine sedimentary rocks.

It was the last item listed in that group, the sedimentary rocks that was of most interest to me. It was the sedimentary Minnelusa formation that was exposed by the Black Hills uplift and eroded the Fairburns into the Badlands.

Although Pabian does not specifically refer to Fairburn agates, he does refer to Teepee Canyon and what he calls South Dakota agates. His chart, which identifies the host rock formation for South Dakota agates, indicates host rock formed during the Pennsylvanian period.

The Minnelusa formation was laid down, in part, during the Pennsylvanian period. This would be consistent with the field observations in the Black Hills relating the agates origin in the Minnelusa limestone.

Pabian describes and distinguishes the formation of what he calls marine sedimentary agates as they would develop under his proposed process of agate formation.

Marine sedimentary agates differ from both thunder eggs and amygdaloidal agates in as much as they crystallize inside masses of amorphous silica or chert. These agates are thought to have formed when sediments became emergent and were weathered and eroded. They do not contain crystalline inclusions or pseudomorphs that characterize amygdaloidal agates. A commonly contained marine invertebrate fossils such as trilobites and brachiopods in Paleozoic examples and clams and rare sand dollars in Myocene-Pliocene sequences.

The spherulitic crystallization of marine sedimentary agates may have been initiated by hematite derived from oxidation of clays or calcite derived directly from limestones. The calcite is confined to the hollow center of the agate.

These kinds of agates have been observed in the middle Pennsylvanian of the Hartville uplift, Wyoming (Harper, 1960), the Black Hills of South Dakota (Chondra and others, 1950), the prior mountains of Montana, where agate cutters call them 'forts', Teepee Canyon agates, and Dryhead agates, respectively. Agatized corals in the tertiary sedimentary sequences of the Tampa Bay, Florida area are attributed to the precipitation of silica in ephemeral, alkaline lakes (Straum, Upchurch, and Rosenzweig, 1981). They consider the silica source to be organic. We do not disagree with their interpretation of the silica source, but we would also consider airfall ash as a possible source. —*page 127*

Theories proposed by both Pabian and Landmesser are plausible given all the field observations that I have had opportunity to make. Further, they are consistent with the geologic history of the Black Hills, which includes repeated marine environments as well as repeated ash falls which have resulted from the volcanic activities both in the immediate area and also in the Rocky Mountains to the west. The gumbo clays of the badlands are the result of those ashfalls.

One of the unique qualities of the finest Fairburns is the formation of colored bands next to a band of clear quartz crystals.

When examining Fairburn agates one becomes aware that there are formations of fairly substantial crystal growths occurring between the agate bands. Layered over the top of these clear crystal areas, one finds a band or bands of agate which appear "feathery". The appearance is caused by the multiple pointed ends of the quartz crystals. The effect can be observed in the Teepee Canyon photo on page 50 and the hills Fairburn photo of an agate from Antelope Spring on page 51.

The possible explanation for this occurrence is the viscosity of the silica gels.

Pabian says:

...the banded portion of the agate nodule is indicative of the stage of growth in which the viscosity of the silica solution was high and the Euhedral quartz crystals represent the timeline in the agate nodule in which the viscosity of a siliceous solution became diluted or low. —*page 20.*

The short space provided for each of the above writers does not do justice to their work. Anyone desiring more technical information must certainly read the original works of these authors.

Historical Postscript—The "New" Good Old Days

Art and Ann Bruce started hunting Fairburns back in the 1940s. They had become addicted to rocks, fossils, Indian artifacts, and other minerals while living in the Black Hills. When they started Fairburn hunting, they were told that it was fruitless because the agates had all been picked up by others.

Each new generation sees the world in a new light. When I began to hunt 30 years later, I was told the same thing. I was disappointed that I had missed the "good old days" when agates could be found.

Myths and tales of large, plentiful, and beautiful agates always abound when rock hunters tell "hunting stories". Why should we be any different than fishermen, deer hunters, or golfers?

Rock hunters tell about the "old days" when hunters used to pick agates by the flour sack-full at State Parks. They claim they could find 15 to 20 Fairburns in one day. They could dig bushels of Teepee Canyons.

It must be remembered that the tales that you hear have come from 75 years of agate hunting. What you now hear is 75 years being compressed into a few minutes. The old-time hunters may now have hundreds or even thousands of agates, but these result from a lifetime of hunting.

There are still wonderful finds being made. The number of finds have diminished over the years because the continued interest in hunting brings new rock hounds to South Dakota hunting areas.

The easy finds in the old days resulted because the badlands was eroding and nobody was hunting for thousands of years. When these agates caught the eye of a rock hound, he had fresh territory to roam in search of Fairburns. If we could close the badlands and grasslands to hunting for a hundred years, hunting would be fantastic again! Since there is only one option, those of us who love Fairburns will have to be satisfied with the limited number that erode out annually or the new finds in the Black Hills.

I have learned over the years, however, that each of us lives in our own "good old days". Wisconsin hunters who hunted in the west back then surely found more agates, but when I look at the old maps, it is also clear that hunting was far more difficult as to travel and access.

My 1941 map of South Dakota indicates that it is 370 miles from Sioux Falls to Rapid City. The old road, Highway 16, was two lanes. It passed through 32 small towns. At that point in time, Rapid City and Sioux Falls were the two largest cities in the state and had a combined total population of less than 50,000 people.

My elderly Wisconsin hunter friends used to drive around the clock in shifts to make Rapid City in a full day. I can now cross Wisconsin, Minnesota, and South Dakota, on safe super highways in less than 13 hours.

The 1941 map shows that the road through the "Big Badlands" (Badlands National Park) was still a dirt road. Mt. Rushmore, which was begun in 1927, was not yet completed. Travel conditions were difficult before the era of convenient and numerous motels and expressways.

You **can** still find agates in the grasslands, badlands, and the Black Hills. With the ease of four-wheel vehicles, the badlands are more accessible. Travel accommodations make today's rock hunting in South Dakota the "new" good old days. South Dakota's huge public national grasslands and national forest literally provide endless hunting areas that reach far into Nebraska. Many new finds are being made and many old areas await someone with the desire for exploration and discovery.

Making an acquaintance or friend in Rapid City, Custer, or Hot Springs, can open many doors to the area and locations not so heavily hunted as are the old Fairburn beds. Every time I have asked to hunt private land, my request has been granted. Rock hunters, just like game hunters, must use common courtesy and ask first for permission to hunt on private land. Take your trash with you when you leave, do not damage fences or roadways and you will likely be welcome to return.

South Dakota has a tremendous interest in tourists. Practically all of the residents will welcome your interest in its natural wonders, not the least of which is the Fairburn agate.

Although Lake Superior agates can be found in Wisconsin, when you tell somebody you're an "agate hunter", they seldom know what that means. In Western South Dakota, they always know what that means. Happy Exploring!

Bibliography

Agnew, Allen F. and Paul C. Tychsen. *A Guide to the Stratigraphy of South Dakota.* Science Center University of South Dakota, Vermillion, South Dakota, 1965.

DeWitt, Ed, J.A. Redden, Anna Burock Wilson, and David Buscher. *Mineral Resource Potential and Geology of the Black Hills National Forest, South Dakota and Wyoming.* U.S. Geological Survey Bulletin No.1580, 1986.

Frazier, Si and Ann. *What's Made of Quartz and Goes "Tweet-Tweet"?* Lapidary Journal, August 1995, p.18.

Gries, John Paul. *Roadside Geology of South Dakota.* Mountain Press Publishing Company, 1996.

Hardson, J.C. and J.R. McDonald. *Guidebook to the Major Cenozoic Deposits of Southwestern South Dakota.* South Dakota Geological Survey Guidebook 2. Science Center University of South Dakota, Vermillion, South Dakota, 1969.

Landmesser, M. *On Boundary P-T Conditions of Agate Genesis.* Chem. Erde 45. Johannes Gutenberg-Universität Mainz, B.R.D., 1986.

_____, *Structure Characteristics of Agates and Their Genetic Significance.* Neues Jahrbuch. Abh.159: 223-235, Stuttgart, Germany, 1988.

_____, *The Problem of Agate Genesis.* Mitt. Pollichia, 72:5-137, Bad Dürkheim/Pfalz, Germany, 1984.

Leiper, Hugh. *The Agates of North America,* Revised Edition. The Lapidary Journal, p.64, 1966.

Madson, John. *South Dakota's Badlands; Castles in Clay.* National Geographic Vol.159, No.4, p.524. April 1981.

O'Harra, Cleophas C. *The White River Badlands.* Department of Geology, Bulletin No. 13, South Dakota School of Mines, 1920.

Pabian, Roger K. and Andrejs Zavins. *Banded Agates Origins and Inclusions.* Educational Circular No. 12. Conservation and Survey Division of the University of Nebraska-Lincoln, 1994.

Pabian, Roger K. *Inclusions in Agates and Their Origins and Significance*: Gems and Gemology, Vol. 16, No. 1, p. 16-28. 1978.

Petsch, Bruno C. and Duncan J. McGregor. *Minerals and Rocks of South Dakota*, Educational Series No. 5. South Dakota Geological Survey, Vermillion, South Dakota, 1973.

Rich, Fredrick J., Editor. *Geology of the Black Hills, South Dakota and Wyoming*, Second Edition. American Geological Institute,1981.

Roberts, Willard L. and George Rapp Jr. *Mineralogy of the Black Hills*, Bulletin No. 18. South Dakota School of Mines and Technology, 1965.

Shaub, Benjamin Martin. *The Origin of Agates, Thunder Eggs, Bruneau Jasper, Septaria and Butterfly Agates*, First Edition. The Agate Publishing Company, 1989.

Zeitner, June Culp. *Midwest Gem Trails*, Third Edition. Published by Gembooks, 1964.

————, *South Dakota: Land of Field Trips*. Lapidary Journal, May 1985, p.248.

————, *Inclusion in Agates and Their Origin and Significance*. Gems and Gemology, Vol. 16, No. 1, p. 16-28.

Notes

Notes